Advances in Aviation Psychology, Volume 2

Since 1981, the biennial International Symposium on Aviation Psychology (ISAP) has been convened for the purposes of (a) presenting the latest research on human performance problems and opportunities within aviation systems, (b) envisioning design solutions that best utilize human capabilities for creating safe and efficient aviation systems, and (c) bringing together scientists, research sponsors, and operators in an effort to bridge the gap between research and applications.

Though rooted in the presentations of the 18th ISAP, held in 2015 in Dayton, Ohio, *Advances in Aviation Psychology* is not simply a collection of selected proceedings papers. Based upon the potential impact of emerging trends, current debates or enduring issues present in their work, select authors were invited to expand upon their work following the benefit of interactions at the symposium. Consequently the volume includes discussion of the most pressing research priorities and the latest scientific and technical priorities for addressing them.

This book is the second in a series of volumes. The aim of each volume is not only to report the latest findings in aviation psychology but also to suggest new directions for advancing the field.

Michael A. Vidulich is a research psychologist in the Air Force Research Laboratory, Ohio. His research interests are aviation psychology, mental workload and situation awareness, and adaptive aiding. He co-edited *Principles and Practice of Aviation Psychology* and *Advances in Aviation Psychology I.*

Pamela S. Tsang is Professor of psychology at Wright State University in Dayton, Ohio. Her research interests are aviation psychology, attention and performance, extralaboratory-developed expertise, and cognitive aging. She co-edited *Principles and Practice of Aviation Psychology* and *Advances in Aviation Psychology I.*

John Flach is Professor of psychology at Wright State University in Dayton, Ohio. His research interests are visual control of locomotion, interface design, decision-making, and motor control. John is co-author of three books and co-editor of two ecological man-machine systems books and an aviation psychology book.

Advances in Aviation Psychology, Volume 2

Using Scientific Methods to Address Practical Human Factors Needs

Edited by Michael A. Vidulich, Pamela S. Tsang, and John Flach

LONDON AND NEW YORK

First published 2017 by Routledge

2 Park Square, Milton Park, Abingdon, Oxfordshire OX14 4RN
52 Vanderbilt Avenue, New York, NY 10017

Routledge is an imprint of the Taylor & Francis Group, an informa business

First issued in paperback 2019

British Library Cataloguing-in-Publication Data
A catalogue record for this book is available from the British Library

Library of Congress Cataloging-in-Publication Data
A catalog record for this book has been requested

ISBN: 978-1-4724-8141-2 (hbk)
ISBN: 978-0-367-88197-9 (pbk)

Typeset in Bembo
by Swales & Willis, Exeter, Devon, UK

In recognition of Raja Parasuraman's important contributions to Aviation Psychology, as both a scientist and a teacher

Contents

Figures

Tables

Contributors

David A. Abbink is an associate professor in the Department of BioMechanical Engineering and leader of the Delft Haptics Lab (www.delfthapticslab.nl). He received his PhD in Mechanical Engineering at Delft University of Technology in December 2006, and was awarded the best PhD dissertation in the area of movement sciences in The Netherlands. His work has been funded by the Dutch National Science Foundation, the Boeing Company (USA), and Nissan Motor Company (Japan). In 2009 and 2015, respectively, he received the prestigious VENI and VIDI personal research from the Dutch National Science Foundation. He is an associate editor for the IEEE Transactions on Human-Machine Systems.

Kasey Ackerman is a PhD student in Mechanical Science and Engineering at University of Illinois at Urbana-Champaign with an interest in aviation flight control systems. He has spent three summers at National Aeronautics and Space Administration (NASA) Langley Research Center in the Dynamic Systems and Control branch (now as part of the NASA Interns, Fellows and Scholars, or NIFS, program) working on advanced control and propulsion systems.

Amy L. Alexander is a member of the Technical Staff at the Massachusetts Institute of Technology Lincoln Laboratory. She specializes in planning and conducting human-in-the-loop (HITL) simulations, primarily in support of Federal Aviation Administration programs. Example HITLs include testing various graphical weather products and alerting mechanisms in the cockpit, exploring the use of a flexible gaming platform to quantify the impact of aviation weather forecast quality on strategic air traffic management decision-making, as well as exploring reduced oceanic separation standards and providing air traffic services to non-towered airports. She received her PhD in Psychology from the University of Illinois at Urbana-Champaign in 2005, and her BS in Psychology from The Ohio State University in 2000. Amy holds a Private Pilot Certification and is a member of the Human Factors and Ergonomics Society and the Association for Aviation Psychology.

Rolf P. Boink received his MS degree in Aerospace Engineering at TU Delft with an interest in human-machine interaction. He is currently a Human

Factors engineer with a Dutch energy distribution grid operator, Alliander NV in the Netherlands.

Dana Broach is a personnel research psychologist at the Federal Aviation Administration's (FAA) Civil Aerospace Medical Institute (CAMI). Dr. Broach's primary research area is personnel selection. His work focuses on the relationship of knowledge, skills, abilities, and other personal characteristics to job and task performance in safety-critical aviation occupations. He has done work in job analysis, human ability taxonomies, test development, job performance measures, test validation, utility analysis, and workforce planning for the air traffic control specialist, electronics technician, systems specialist, and federal air marshal occupations in the FAA. Dr. Broach has published over 100 journal articles, book chapters, technical publications, and conference presentations. He is a member of the Society for Industrial and Organizational Psychology, Association for Psychological Science, the Association for Aviation Psychology, the European Association for Aviation Psychology, the Australian Association for Aviation Psychology, and the American Institute for Aeronautics and Astronautics. He earned his master's degree and doctorate in Industrial/Organizational Psychology from the University of Tulsa in 1991 and has worked at CAMI since 1989.

Cristina L. Byrne has worked as a personnel research psychologist for the Federal Aviation Administration's (FAA) Civil Aerospace Medical Institute in the Human Factors Research Division since 2011. Dr. Byrne's primary research relates to the selection, training, and placement of air traffic controllers. Her current work focuses on understanding the personnel and organizational factors that are related to the successful completion of the FAA's rigorous and complex air traffic controller training program. Dr. Byrne has done work in training and organizational development, survey and test development, validation, and utility analysis for the air traffic control workforce. She is a member of the Society for Industrial and Organizational Psychology, as well as the Human Factors and Ergonomics Society. Dr. Byrne received her PhD in Industrial and Organizational Psychology from the University of Oklahoma.

Ronald Carbonari is a research programmer with the Beckman Institute at University of Illinois at Urbana-Champaign with an interest in human-in-the-loop simulation. He has contributed to aviation safety research and education for more than 25 years.

James P. Chamberlain is founder and president of Sunrise Aviation, Inc. and aerospace research consultant after a 32-year career at National Aeronautics and Space Administration Langley Research Center. His research interests include flight crew systems and airspace operations such as integration of unmanned aircraft systems into the national airspace. He earned a BS and MS in Mechanical and Aerospace Engineering, respectively, from the University of Florida and is a 3,700-hour commercial pilot with experience in numerous aircraft types from gliders to business jets.

Sebastian Clauß received his BSc and MSc degrees in Aerospace Engineering from the University of the Bundeswehr Munich (UBM) in 2009 and 2011, respectively. From 2011 to 2014 he served as research assistant at the Institute of Flight Systems, UBM. He is shortly to defend his thesis to become a doctor of engineering. His research focuses on human-agent interaction within highly automated uninhabited aerial vehicles systems and the adaption of established concepts of human-human interaction therein. Sebastian Clauß holds the rank of captain in the German Air Force.

James R. Comstock, Jr. is a Research Aerospace Engineer at the National Aeronautics and Space Administration Langley Research Center in Hampton, Virginia. He received a PhD in Industrial/Organizational Psychology (Engineering and Systems) from Old Dominion University, Norfolk, Virginia, in 1983. His research interests include visual performance and perception applied to display design and interface issues; performance assessment in high technology environments; and issues of human-automation interaction and related interfaces.

Maria C. Consiglio is a Senior Research Scientist at National Aeronautics and Space Administration Langley Research Center in Hampton, Virginia. She has a Bachelor degree in Scientific Computation from the University of Buenos Aires and a Masters degree in Computer Science from the University of Pittsburgh. Her work in air traffic management research spans more than 15 years and includes over 30 publications including work on autonomous operations and separation assurance technologies, and more recently on unmanned aircraft systems integration and detect-and-avoid research. She has served as a Project Engineer for the NASA Unmanned Aerial System integration in the NAS project.

Joseph T. Coyne is a Research Psychologist and head of the Warfighter Human Systems Integration Laboratory at Naval Research Laboratory. Dr. Coyne received his PhD in Human Factors Psychology from Old Dominion University in 2004. His graduate work looked at pilot decision-making in inclement weather conditions and assessed how a new display altered pilot decisions. His graduate work including his dissertation was funded under several National Aeronautics and Space Administration Fellowships and internships. Dr. Coyne has served as the principal investigator on a number of diverse research efforts focusing on human performance and physiological assessment of the warfighter. He has worked directly with a number of user communities such as the Marine Corps, Army, and Law Enforcement. While working with the Navy he has investigated the effectiveness of using physiological assessment as a tool to measure changing skill level during a simulated uninhabited aerial vehicle laboratory task. He was the principal investigator in a study measuring the impact of a motion mitigating display in a command and control vehicle, he evaluated virtual environments for infantry simulation, and has conducted several cognitive task analyses to better understand the needs of the warfighter.

Mag. rer. nat Elisabeth Denk received her university diploma degree in Psychology from the University of Graz, Austria in 2012. Between 2013 and 2015 she was research assistant at the Institute of Flight Systems of the University of the Bundeswehr Munich. Her research interests focus on pilot-in-the-loop experimentation to establish behavior models based on adaptive strategies, under various workload conditions. She uses eye-tracking and behavioral methods to investigate the continuous impact of workload in military cockpits.

Col. John R. Dougherty trained as a USAF pilot and flew the F-16 for 16 years and the MQ-1 Predator Remotely Piloted Aircraft (RPA) for 8 years. He received his Bachelor of Science degree in Electrical Engineering/ Computer Option from North Dakota State University. Col. Dougherty has served in numerous squadron positions including Chief of Training, Weapons, Scheduling, and Wing Safety. In addition, he has served in numerous leadership positions culminating in his present role as the Operations Group Commander for the 119th Operations Group. He commands three squadrons with over 250 Airmen and is responsible for all flying and RPA combat operations. Col. Dougherty is a Command Pilot with more than 3,500 hours in 8 aircraft.

William (Bill) R. Ercoline, Lt, Col, (retired), USAF, has an MS Engineering Physics degree from Air Force Institute of Technology and a PhD in Engineering Management. He has over 4,000 flight hours in a variety of military and general aviation aircraft, and more than 30 years of research experience in the areas of human factors, spatial disorientation, flight symbology development, and general aviation psychology and physiology. He is a former Associate Professor of Physics at the USAF Academy. He currently manages the San Antonio centrifuge and altitude chambers operation for the WYLE Science, Technology & Engineering Group at Brooks City-Base, Texas (formerly Brooks AFB). His team of researchers and technicians provide life support equipment research, development, testing, and evaluation to the 711 Human Performance Wing and the USAF School of Aerospace Medicine, now located at Wright-Paterson Air Force Base, Ohio. He consults with USAF accident investigation boards if spatial disorientation issues are a factor, and he provides lectures for several of the USAF School of Aerospace Medicine education programs. He has published many articles about the costs, causes, and countermeasures of spatial disorientation, and co-edited and co-authored with Dr. Fred Previc, the textbook, *Spatial Disorientation in Aviation*. Bill lectures internationally on the subject of spatial disorientation, and serves on multi-service working groups and international organizations specializing in aircrew performance in high workload environments. He has provided laser eye protection research support for the Directed Energy Branch of the Human Effectiveness Directorate within the Air Force Research Laboratory. He has served as an Aerospace Medicine Association (AsMA) journal paper reviewer, is a fellow of the organization, and has been an officer for several of the sub-organizations within AsMA;

he has presented several papers at the International Association of Aviation Psychology Symposiums. He and his wife Kathy have been married for 47 years and they have two grown children and three grandchildren.

Capt. Karl E. Fennell has been an airline pilot for 21 years and is currently flying Captain on the Boeing 737NG. He also has experience in human factors research and flight safety. Capt. Fennell earned his MA in cognitive psychology from The University of Denver in 1993 and his BA in psychology from Miami University in 1988. He is also serving in the Air Force Reserves in global contingency operations at the 618th Air Operation Center. In the Air Force, he was formerly flying the Lockheed C-130. Capt. Fennell was the Director of Human Factors for the Air Line Pilots Association from 2008 to 2011. He was also part of the B787 training development team to improve content and delivery in 2006 to 2007. Capt. Fennell has worked on several human factors projects as subject matter expert and collaborator and has specialized in automation training and development. His research interests lie in developing and applying solutions to improve aviation safety and overall experience with flight technology.

John Flach received his PhD (Human Experimental Psychology) from the Ohio State University in 1984. John was an assistant professor at the University of Illinois from 1984 to 1990 where he held joint appointments in the Department of Mechanical & Industrial Engineering, the Psychology Department, and the Institute of Aviation. In 1990 he joined the Psychology Department at Wright State University. He served as department chair from 2004–2013. He currently holds the rank of professor. He teaches graduate and undergraduate courses in the areas of experimental cognitive psychology and human factors. John is interested in general issues of coordination and control in cognitive systems. Specific research topics have included visual control of locomotion, interface design, decision-making, and motor control. John is particularly interested in applications of this research in the domains of aviation, medicine, highway safety, and assistive technologies. John is co-author of three books *Control Theory for Humans* (with Rich Jagacinski), *Display and Interface Design* (with Kevin Bennett), and *What Matters?* (with Fred Vorhoorst). He is also a co-editor of two books on ecological approaches to man-machine systems and an earlier volume on aviation psychology.

Scott Galster is the Chief of the Applied Neuroscience Branch, Warfighter Interface Division, Human Effectiveness Directorate, 711 Human Performance Wing at Wright-Patterson Air Force Base. Dr. Galster is accountable for strategic planning, development, execution, and reporting of a diverse body of research and development efforts examining the applied neuroscience aspects of decision-making and performance spanning the identification of genetic factors influencing stress responses to increasing distributed team performance. Dr. Galster is the founder of the Sense-Assess-Augment paradigm used across the Department of Defense to sense operator state, assess that state relative to performance, and provide focused and personalized augmentations to assure mission success. Dr. Galster holds a PhD in Applied Experimental Psychology from The Catholic University of America in Washington, DC. He is the recipient of a NASA fellowship award, the Paul Fitts Engineering award, and the International Program award for Human Effectiveness. Dr. Galster is also an Adjunct Professor, College of Medicine, Department of Neuroscience and the College of Engineering, Biomedical Engineering Department at the Ohio State University.

Rania W. Ghatas is a human/machine systems research engineer at National Aeronautics and Space Administration Langley Research Center in Hampton, Virginia. She obtained her MS degree in Human Factors and Systems Engineering in 2011 at Embry-Riddle Aeronautical University in Daytona Beach, Florida and earned her BS degrees in Molecular Biology and Microbiology, and in Psychology in 2009 at the University of Central Florida.

Christina Gruenwald obtained her BS degree in Psychology from Wittenberg University in 2015. She is currently a research associate in the Human Universal Measurement and Assessment Network (HUMAN) Laboratory contracted through Oak Ridge Institute for Science and Education. She has been working in the HUMAN Lab since November 2014. The HUMAN Lab is a part of the Applied Neuroscience Branch (AFRL711 HPW/RHCP) located at Wright-Patterson Air Force Base. The HUMAN Lab has adopted the Sense-Assess-Augment paradigm as its framework of research. The focus is to sense RPA operators cognitive state using physiological measures, assess their workload using models and algorithms that detect when they might be overloaded, and finally, to apply some source of augmentation in order to reduce workload and improve performance. In the HUMAN Lab, she has assisted with experimental design, physiological sensor application, data collection, and data analysis. In addition to supporting Sense-Assess-Augment research, Christina has been involved in the research, validation, and improvement of eye metrics used within the HUMAN Lab. Prior to her work in the HUMAN Lab, Christina worked as a research assistant to Dr. Jeffrey Brookings at Wittenberg University where she researched predictors of academic achievement, anthropomorphism, and rejection sensitivity.

Hans J. Hoermann is a Senior Human Factors Scientist at the German Aerospace Center in Hamburg. He received his PhD in Applied Psychology from Free University in Berlin in 1987 and has accumulated 30 years of research experience on aviation human factors. His research interests cover human performance and fatigue of flight crewmembers, cockpit resource management training, and pilot aptitude testing. Hans has published over 100 research papers. The most recent ones address the impact of present and future levels of automation on the human role in the cockpit. In collaboration with airline training organizations, he developed several training courses specifically for crew resource management and flight instruction. As a Technical Fellow for Safety and Human Factors, Hans spent more than three years at Boeing's Research & Technology Center in Madrid. He was awarded a fellowship in the Royal Aeronautical Society in 2006 and is a registered aviation psychologist in the European Association for Aviation Psychology. Hans serves as domain editor for the *International Journal of Aviation Psychology*. In his free time he enjoys flying single engine airplanes as a private pilot.

Keith D. Hoffler is President, Senior Research and Development Engineer, and founder of Adaptive Aerospace Group in Hampton, Virginia. He received BS and MS degrees in Aerospace Engineering from North Carolina State University. He has over 30 years of experience in aerospace research and development. Keith holds a Commercial/Instrument Rating and flies the company's Cessna R182 for business travel and to evaluate various systems.

Naira Hovakimyan received her PhD in Physics and Mathematics in 1992 in Moscow from the Institute of Applied Mathematics of the Russian Academy of Sciences, majoring in optimal control and differential games. She is currently W. Grafton and Lillian B. Wilkins Professor of Mechanical Science and Engineering with an interest in robust, adaptive control systems in aviation and other applications. She is recipient of the 2015 Society of Women Engineers Achievement Award, "For advancing a new control methodology with far-reaching applications and for pioneering contributions to the field of robust adaptive controls in aeronautics and aviation."

Alex Kirlik is Professor of Computer Science and Member of the Beckman Institute at University of Illinois at Urbana-Champaign with an interest in human factors, cognitive science and engineering, and human-computer interaction. He received his PhD in Industrial & Systems Engineering (human-machine systems) from The Ohio State University. In addition to Illinois, he has held permanent or visiting positions at Georgia Tech, Stanford University, National Aeronautics and Space Administration Ames Research Center, Yale University, the University of Connecticut, and the Liberty Mutual Research Institute for Safety. He is known for his contributions to human-automation interaction and modeling human judgment, situation awareness, and skilled performance, and is co-editor (with John D. Lee) of the *Oxford Handbook on Cognitive Engineering*, and is editor of the Oxford Series on Human-Technology Interaction.

Robert Mauro conducts basic and applied research in decision-making, risk assessment, and human emotion. His work includes laboratory and field work on pilot decision-making, training, procedure development, and automation. He has also worked on the development of risk assessment tools and on decision-making in space flight control. Dr. Mauro has worked with National Aeronautics and Space Administration, the Federal Aviation Administration, the National Science Foundation, major airlines, and other government agencies and private companies. His research utilizes experimental, survey, and observational methods, and psychological and physiological measures. Dr. Mauro received an AB from Stanford University in 1979, an MS from Yale University in 1981, and a PhD from Stanford in 1984. He is currently on the faculty at the University of Oregon where he teaches courses in human factors and statistics. He serves on the Executive Committee of the Institute of Cognitive & Decision Sciences and is Senior Research Scientist at the independent Decision Science Research Institute.

Jason S. McCarley holds a PhD in Experimental Psychology from the University of Louisville. He held postdocs at the Naval Postgraduate School and the University of Illinois, and then faculty positions at Mississippi State University and the University of Illinois. He is currently Professor in the School of Psychology at Flinders University of South Australia, and has been associate editor of *Journal of Experimental Psychology: Applied* since 2012. His lab uses behavioral methods and mathematical modeling to study basic and applied aspects of attention and human performance. Specific topics of interest include visual search, driver distraction, and human-automation interaction.

Matthew Middendorf is currently the laboratory manager for the Human Universal Measurement and Analysis Network (HUMAN) Laboratory located at Wright-Patterson Air Force Base, Dayton, Ohio. This laboratory is in the Applied Neuroscience Branch (AFRL 711 HPW/RHCP). Mr. Middendorf has 36 years of experience conducting experimental research with a focus on real-time programming, algorithm development, and signal processing. In his current position, Mr. Middendorf has developed algorithms to extract features from physiological signals, which can be used for cognitive state assessment. This work supports the Sense-Assess-Augment paradigm used to optimize human operator performance in remotely piloted aircraft operations. In prior research, he has developed software to allow hands-free activation of icons using steady-state visually evoked responses based on numerical processing of EEG signals. Mr. Middendorf was the lead engineer on a research project to develop a force cueing evaluation methodology using a dynamic seat, G-suit, and combat edge. He also has extensive knowledge in all areas of flight simulation, including turbulence modeling, control scaling, flight control systems, aerodynamics, cockpit instrumentation, motion platforms, and out-the-window scene generation.

Samuel S. Monfort received an MA in Experimental Psychology from Wake Forest in 2012, and is currently a PhD student in Human Factors and Applied Cognition at George Mason University. His research at George Mason is funded by the Office of Naval Research and is focused on evaluating immersive infantry training systems. He also works part-time at Perceptronics Solutions, Inc. as a junior human factors scientist and at the Naval Research Lab as a research assistant. His research interests include human performance modeling, trust in automation, statistics, and research design.

Mark Mulder received his MSc degree in Aerospace Engineering from Delft University of Technology in September 2000 and his PhD degree on Haptic Driver Support Systems in January 2007. From 2008 to 2014 he studied Medicine and obtained his medical degree in December 2014. Until then, he also continued to work as a post-doctoral research fellow at the Biomechanical and Aerospace Engineering Departments of Delft University of Technology. Currently he is a medical doctor working in internal medicine.

Max Mulder received the MSc and PhD (cum laude) degrees in aerospace engineering from the Delft University of Technology, Delft, The Netherlands, in 1992 and 1999, respectively, for his work on the cybernetics of tunnel-in-the-sky displays. He is currently a Full Professor and Head of the Control and Simulation Section, Faculty of Aerospace Engineering, Delft University of Technology. His research interests include cybernetics and its use in modeling human perception and performance, and cognitive systems engineering and its application in the design of ecological human-machine systems.

Esa M. Rantanen trained as a commercial pilot. He also has seven years of experience as an air traffic controller and an air traffic control instructor. Dr. Rantanen has a BS and a Master of Aeronautical Science degree from Embry-Riddle Aeronautical University, Daytona Beach, Florida. He also has an MS in Industrial Engineering degree from the Pennsylvania State University, with specialization in human factors/ergonomics engineering. His PhD degree is from Penn State as well, in Engineering Psychology. Dr. Rantanen has served as an Assistant Professor at the Institute of Aviation of the University of Illinois at Urbana-Champaign. Presently, Dr. Rantanen is an Associate Professor of Psychology at the Rochester Institute of Technology. He is primarily teaching courses in the MS in Experimental Psychology program and supervising graduate students' thesis research in the Engineering Psychology track of the program. Dr. Rantanen's research interests lie in the areas of human factors in complex systems, human performance measurement and modeling, mental workload, decision-making, and human error and reliability.

Kenyon Riddle received his MS degree in Human Factors at University of Illinois at Urbana-Champaign with an interest in human-automation interaction. He is currently a Human Factors Scientist with Aptima, Inc., Orlando, Florida.

MSgt. Jason Russi is a 16-year veteran of the U.S. Air Force and is currently the Tower Chief Controller at Wright-Patterson Air Force base, Dayton, Ohio. MSgt. Russi has operated as an air traffic controller in four states and three countries and is also a human factors researcher with a patent pending for stereoscopic radar display technology. He holds MS degrees in Aeronautical Science from Embry-Riddle Aeronautical University and Information Resource Management from the Air Force Institute of Technology where he is currently enrolled as a PhD student in the Systems Engineering program.

Axel Schulte received his university diploma degree in Aerospace Engineering (focus in control engineering) in 1990 and his Doctor of Engineering (focus in aviation human factors) in 1996, both from the University of the Bundeswehr Munich (UBM). From 1995 to 2002 he worked for the aviation industry as a systems engineer and project manager in several research and technology projects in the fields of pilot assistant systems, mission management systems, and cockpit avionics of military aircraft. Since 2002, he has been Full Professor of Aircraft Dynamics and Flight Guidance at the Aerospace Engineering Department of UBM and Head of the Institute of Flight Systems. His research interests are in the areas of cognitive and cooperative automation in flight and military mission management, and in human-automation integration in aviation. Axel Schulte was Visiting Professor at the Humans and Automation Laboratory of Massachusetts Institute of Technology in 2010.

Benjamin Seefeldt was an MS student in the Department of Computer Science at University of Illinois at Urbana-Champaign with an interest in human-computer interaction and human-automation interaction. His other interests include data visualizations, the effects of technology on interpersonal relationships, and machine learning. He recently completed a position as User Experience developer/researcher at John Deere.

Lui Sha received his PhD from CMU in 1985 and joined University of Illinois at Urbana-Champaign in 1988 where he is Donald B. Gillies Professor of Computer Science with an interest in software engineering and verification among related topics. He is a Fellow of IEEE and recipient of the 2016 IEEE Simon Ramo Medal, "for technical leadership and contributions to fundamental theory, practice and standardization for engineering of real time-systems." His work on real-time and safety-critical system integration has impacted many large-scale, high technology programs, including GPS, Space Station, and Mars Pathfinder.

Ciara M. Sibley is an Engineering Research Psychologist in the Warfighter Human Systems Integration Laboratory at the U.S. Naval Research Lab. The primary focus of her research has been within workload management and the use of neurophysiological data to infer user state and predict task performance. In addition, she has conducted several cognitive task analyses with operational communities, including Unmanned Aerial System (UAS) operators and law enforcement. Ms. Sibley has contributed to multiple research efforts within the unmanned systems domain and is currently the principal investigator on a project aimed at developing a dynamic task scheduler for UAS operators. Ms. Sibley has also engaged in program management supporting multiple programs at both the Defense Advanced Research Projects Agency and Office of Naval Research. She currently serves as the Deputy Program Officer for the Command Decision Making program, helping to oversee a new portfolio of research focusing on the development of proactive decision support technologies.

Donald Talleur has an MS in Psychology from the University of Illinois and is an experienced pilot, currently employed as an assistant chief flight instructor with the Parkland College Institute of Aviation. He has contributed to aviation safety research and education for more than 25 years.

Julia Trippe is a doctoral student in Linguistics at the University of Oregon, a research assistant in human factors aviation at Decision Research, and a certified flight instructor. She investigates pilots' understanding of aircraft automation systems in an attempt to bridge the gap between avionics engineers and airplane pilots. Julia also investigates Aviation English speech and comprehension as a means of establishing fair and effective standards for international pilots and controllers. She has been a pilot since 1994; has taught pilots and instructors; and flown for survey, cargo, commercial, and corporate operators.

Pamela S. Tsang is Professor of Psychology at Wright State University in Dayton, Ohio. She received her AB from Mount Holyoke College and her PhD from the University of Illinois at Urbana-Champaign. Previously, she was a National Research Council post-doctoral fellow at National Aeronautics and Space Administration Ames Research Center. Her research interests are attention and performance, extralaboratory-developed expertise, cognitive aging, and aviation psychology. She is interested in applications of her research in a wide variety of domains that include aviation, surface transportation, and medicine. She co-edited the volume, *Principles and Practice of Aviation Psychology* with Michael Vidulich.

M. M. (René) Van Paassen received his MSc degree (cum laude) from the Delft University of Technology, Delft, The Netherlands, in 1988, and a PhD in 1994, both on studies into the neuromuscular system of the pilot's arm. Thereafter, he was a Brite/EuRam Research Fellow with the University of Kassel, Germany, where he worked on means-ends visualization of process control dynamics, and a post doc at the Technical University of Denmark. René is currently an Associate Professor in Aerospace Engineering at the Delft University of Technology, working on human-machine interaction and aircraft simulation. His work on human-machine interaction ranges from studies of perceptual processes, haptics and haptic interfaces, and human manual control to design of and interaction with complex cognitive systems. René is a senior member of IEEE and of AIAA, and an associate editor for IEEE Transactions on Human-Machine System.

Michael A. Vidulich is a Senior Scientist at the Air Force Research Laboratory's Human Effectiveness Directorate's Applied Neuroscience Branch. He served as the Technical Advisor for the Warfighter Interface Division from 2006 to 2013. He is also a member of the adjunct faculty of the Wright State University Department of Psychology, where he has taught since 1989. Previously, he was a research psychologist at National Aeronautics and Space Administration Ames Research Center. He received a BA in Psychology from the State University College of New York at Potsdam, an MA in Psychology from The Ohio State University, and a PhD from the University of Illinois at Urbana-Champaign. His research specializes in cognitive metrics for human-machine interface evaluation and adaptation. He co-edited the volume, *Principles and Practice of Aviation Psychology* with Pamela Tsang, and the volume *Advances in Aviation Psychology – Volume 1* with Pamela Tsang and John Flach.

Michael J. Vincent is a Research Engineer at National Aeronautics and Space Administration Langley Research Center in Hampton, Virginia. He earned his BA in Psychology from Wichita State University and his MS in Human Factors from Embry-Riddle Aeronautical University. His work has primarily been in cockpit display and interface design and evaluation. He currently holds a private pilot certificate.

Christopher D. Wickens is a Professor of Psychology at Colorado State University, a Senior Scientist at AlionScience Company in Boulder, Colorado, and Professor Emeritus of Psychology and Aviation at the University of Illinois, where he headed the Aviation Human Factors Division from 1985 to 2005. His primary research interests are in human attention, aviation displays, and human-automation interaction. In particular, he has developed cognitive models of visual attention and supervision, multi-tasking, and workload which have been deployed in the study of human interaction with imperfect automaton.

Enric Xargay graduated with a PhD in Aerospace Engineering and then become a Postdoctoral Research Associate at University of Illinois at Urbana-Champaign with an interest in aviation flight control systems. During his graduate work he was recipient of the Roger A. Strehlow Memorial Award from the Department of Aerospace Engineering in recognition of outstanding research accomplishment. He is currently Co-Founder and Director of a startup that focuses on the development of control technologies for autonomous systems.

Part I

Aviation psychology

1 The history of instrument flight

(As told to me by Carl J. Crane)

William R. Ercoline

Prologue

> The evolution of man saw him develop over millions of years as an aquatic, terrestrial, and even arboreal creature, but never an aerial one. In this development, he subjected himself to and was subjected to many varieties of *transient* motions, but not to the relatively *sustained* linear and angular accelerations commonly experienced in aviation. As a result, man acquired sensory systems well suited for maneuvering under his own power on the surface of the earth but poorly suited for flying . . . it should come as no surprise that his sudden entry into the aerial environment resulted in a mismatch between the orientational demands of the new environment and his innate ability to orient. The manifestation of this mismatch is spatial disorientation.
>
> (Gillingham & Previc, 1996, p. 339–340)

The year is now 2016. Kent penned those words about thirty years ago. It is still amazing (and troubling) that spatial disorientation remains arguably the most significant human factors issue in aviation today (Gibb, Ercoline, & Scharff, 2011, p. 3). Kent wrote, “it should come as no surprise,” yet special disorientation still does come as a surprise. Pilots are surprised by it all the time. Many people still believe and assume natural instincts are better than flight instruments. If the attitude indicator doesn’t look quite right, tap on the case to fix it. One’s perception can’t be wrong . . . the instrument must be hung up. That’s usually the first reaction when a disagreement exists between a pilot’s sense of orientation and his or her artificial horizon instrument, also known as the attitude indicator.

We’ve known about the fallacy of our orientation senses for almost ninety years, yet we still have trouble dealing with them when confronted with a sensory conflict in flight. How many lives would have been saved had we considered this more of a problem early in our quest to fly? Change has occurred, but the process has been slow, and the change is still going on today. How we first came to recognize the problem and to start believing in the airplane’s flight instruments instead of one’s perceived orientation, is a story worthy of a closer look. There were many involved in this process, but at the start there were only a handful of people who braved to go against accepted opinion and chose to make a difference.

Figure 1.1 Carl Crane and Bill Ercoline in Hangar 9, Brooks AFB, TX.
Source: Personal photo.

The story I'm about to share with you was first shared with me in the early 1970s by Carl J. Crane (see Figure 1.1). He told me about a discovery made around 1926 by a fellow pilot and friend of his. The discovery would begin to change forever the way we flew and would fly airplanes. The exact change itself, and our understanding of the need for change, took time. It didn't happen quickly, but looking back on it from today, it happened over a span of about twenty-five years. And when you take into account how important this discovery was to the overall success of man's ability to fly an airplane safely, it is arguably the most significant scientific finding of the twentieth century. And as many brilliant discoveries go, the discoverer never got to reap the benefits, and the recognition he did receive, in large part he could have done without.

Unless you are a real military history buff, it's quite likely you have never heard of his name. He was a military pilot (one of the first 300 to be licensed), and his insistence on the need for instrument flight changed everything. His story, this story, is for all of us—those who have flown and those who fly. It is our legacy.

Much of this story has been supported by historical research, but not all. Its accuracy is principally based upon the recollection of the person who lived it—Carl J. Crane. And it's documented here for those who have an interest in knowing more about the way it once was. And as it was told to me by Carl, it's more than just a story of instrument flight. It's also a story about human nature and human factors. But since Carl first called it "The history of instrument flight," we'll leave it as such. This chapter is dedicated to Carl's memory, his many accomplishments, and to the memory of his colleague and friend William (Bill) C. Ocker. Carl felt Bill never received the recognition deserving of his discovery. Make no mistake about it; their work and the results from their work together made the world of aviation a safer place for all of us.

John Macready

In 1923, an aviation record was established by Lt.s John Macready and Oakley Kelly. They were the first to successfully fly an airplane non-stop across the United States (see Figures 1.2 and 1.3). They tried several times earlier, without

Figure 1.2 T-2 Fokker in flight.

Source: Image courtesy the National Museum of the USAF®.

Figure 1.3 Lts Macready and Kelly.
Source: Image courtesy the National Museum of the USAF®.

success, but in the early days of May, 1923, they were able to take off from New York on May 2 and land in San Diego on May 3. They would eventually be awarded a McKay Trophy for this remarkable accomplishment (Macready would go on to be the only person to ever win three), and the flight was considered by many to be one of the most significant airplane accomplishments at that time. The flight clearly demonstrated that it was possible to travel safely from one side of the U.S. to the other in about a day without stopping. It was such an accomplishment National Geographic devoted an entire issue to the flight in the summer of the following year (*The National Geographic*, July 1924). They asked John to write up the story, and he obliged.

One of the challenges with a flight of such magnitude was with the amount of time needed to sustain the flight over the lengthy distance. And because of the long distance and slow speed of the airplane, the flight would have to take place overnight. A significant amount of ground support was also needed to help with navigation, and the entire story itself is worth reading. But it's not the successful accomplishment of John Macready and Oakley Kelly that is of interest to the story of the history of instrument flight. It's something John wrote about describing his thinking and understanding that causes us to reflect on the flying situation at the time.

On a page of the *National Geographic* publication, John Macready (who wrote a large portion of the article) makes an interesting statement. Most pilots would not have made the statement, because in 1923 pilots believed they had an innate skill that allowed them to maintain the spatial orientation of the airplane—the ability to keep the airplane upright by the seat of their pants. They didn't need instruments to do it! (Too many pilots still believe that today.) They possessed this natural skill to fly. This innate ability was their earned rite of passage. After all, they flew on a regular basis and they earned their wings because they had the right kind of stuff. Macready states in the publication, "I, personally, believe that if there is no fixed outside point or horizon, no one can tell his position, whether upside down, straight up, or crosswise, except when the force of gravity pulls him away from or toward the plane" (*National Geographic*, July 1924, p. 68). John wrote that statement for all to read.

Considering his words went against popular thinking, it took some courage to make the statement, but John was that type of pilot. He stated what he found to be true and not worry what others believed, even if it was contrary to current thinking. As mentioned earlier, most pilots at this time believed they could fly by the "seat of their pants." They had a natural skill for flying, and, as you would expect, they were very proud and protective to have such a skill.

John knew something wasn't quite right with the situation. He realized, most likely through his own flight experience, that one's sense of orientation required him to see something outside the airplane. And he considered his opinion important enough to write it up, and this simple statement could very well be the first such statement for an American pilot to make. Others may have known about the problem, but John was the first to write it up.

Let's go a little further into Macready the test pilot and the situation. As you may know, there were many early flying challenges during the 1920s. A lot of the early flight testing was done out of McCook Field (Dayton, Ohio). John Macready happened to be the chief test pilot of a small group of test pilots. Figure 1.4 is a photo of John standing in the middle of the back row. Jimmy Doolittle happens to be in the front row, third from the right. (He'll become a player in the history of instrument flight as well.) Many of these pilots were setting altitude records, speed records, and distance records. It was a time "we" were learning what it meant to fly. As Carl explained it—learn we did! One of the new things these test pilots were experiencing on a regular basis was flight during times of low visibility (obscure weather) and at night. It was an environment not many pilots had any experience with.

Because they were test pilots, and experiencing firsthand what John Macready described in his transcontinental story, they began to develop techniques that would help them when they flew for periods of time without seeing outside the cockpit. One such technique they developed was to shake your head and move your body around. Do it just before going into clouds and this would "cage the gyros" in your head. Another technique developed was to put the control stick between your knees and not touch it until you were clear of the clouds. This technique probably worked as long as you didn't have

Figure 1.4 Early test pilots of the U.S. Air Army Service, 1920s at McCook Field, Dayton, Ohio.

Source: Image courtesy the National Museum of the USAF®.

Note: From left, back row: Unknown; Lt. Harold A. Johnson; Lt. E. Hoy Barksdale; Lt. John A. Macready, chief test pilot; Lt. James T. Hutchinson; Lt. Reuben C. Moffat; Louis G. Meister, civilian test pilot; and Unknown. From left, kneeling, Unknown; R. G. Lockwood, civilian test pilot; Lt. George Tourtellot; Lt. James Doolittle; Lt. William N. Amis; and Unknown.

to change heading while in the clouds, and it probably wasn't very helpful for night flying. But it was a technique used to help mitigate a problem several pilots had experienced. There were other techniques, but it's obvious from these two that they were grasping for something to help them keep the airplane upright. It's also pretty safe to say, based on Carl's personal experience and the reports given by many of these early pilots, they didn't share their problems with each other. It would have been viewed as a weakness.

The orientation problem wasn't just caused by the psyche or macho image of the pilots. It is instructional to look at some of the airplane flight instruments of those early cockpits. It appears the attitude of the pilots transcended into the design and maintenance of the instruments in the cockpits. Pilots were adamant they did not need instruments to fly. Figure 1.5 is a depiction of one of the premier airplanes of the decade of the 1920s, the Curtiss P-1C Hawk—an airplane that was flown throughout the world. The airplane itself is not the

Figure 1.5 Curtiss P-1C "Hawk" on the "flight line."
Source: Image courtesy the National Museum of the USAF®.

problem. It's the instruments in the cockpit that shed light on the problem. A photo of the cockpit of the airplane can be seen in Figure 1.6. Note the lubber lines that were added to the cockpit photo. They were inserted by Carl to point out the various instruments and gauges. Not all airplanes had all the same instruments, but this one had something rare at the time. You can easily see the altimeter and some other engine instruments, and the lubber lines leading to them, but it also contained a Bank and Turn Indicator and whisky compass which were carefully hidden. Search the photo and follow the lubber lines that show the locations of the Compass and the Bank and Turn Indicator. If you find the lubber lines (you won't find all the words of these two instruments), follow them to their end points and you'll see the lines end up at the bottom of the photo. The Bank and Turn Indicator and the Compass were mounted on the floor of the cockpit—between the rudder pedals and behind the stick! Carl explained that pilots claimed they didn't need these kinds of instruments so they were installed out of the way. Should the pilot want to use them, he would have to move the stick to the side in order to see either the Compass or the Bank and Turn Indicator.

To make matters even a little more difficult, about this same time personnel within the medical corps (flight surgeons) were working on the physical

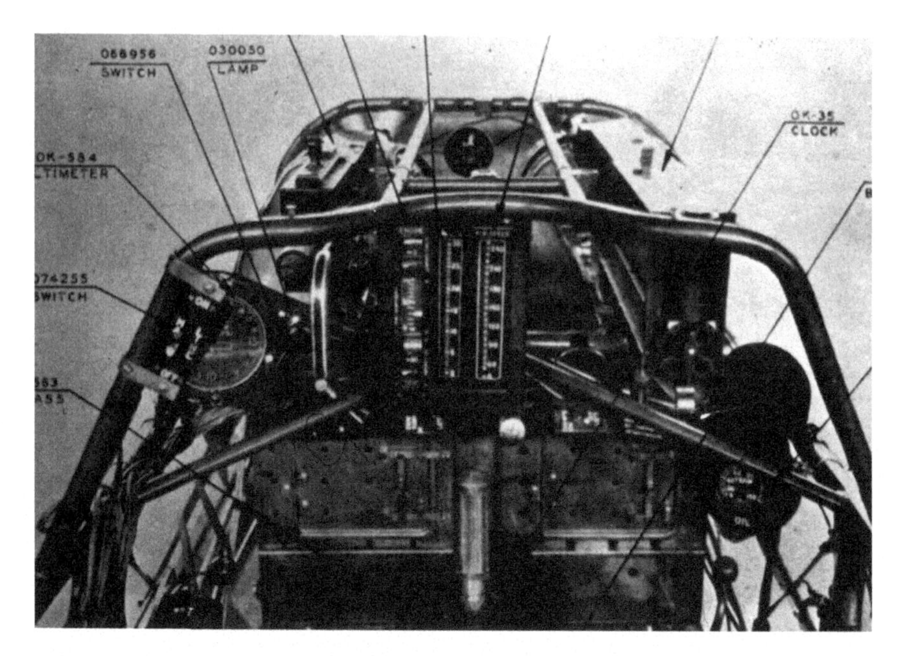

Figure 1.6 Instrumented cockpit of the Curtiss P-1C.
Source: Image courtesy Carl Crane.

tests needed to determine if a pilot was healthy enough to fly (Noe, 1989, p. 80–99). With some support from pilots, the doctors were busy determining physical standards that could be used to determine a pilot's physical condition for flight. Eye tests, fitness standards, and a lot of other tests were useful. One that was really considered important was the health of the vestibular apparatus (the inner ear). The thinking was that a pilot needed a healthy vestibular apparatus when in flight, especially when he couldn't see outside the cockpit. They were correct in a way, the pilot did need a healthy vestibular system; but by the early 1920s they went so far as to make the vestibular testing more important than the visual system. Rotating chairs and multi-axis rotational seats were designed to help identify those who had the natural skill required to fly. (Even after the discovery of the need to use instruments for orientation, it would take years to change.)

To sum up the situation, airplane flight had evolved within a few decades into a troublesome condition. Pilots "believed" they didn't need instruments to fly the airplanes, manufacturers weren't really sure where to put what few instruments a pilot would need, and the medical community was focused on a good sense of balance for pilot selection. A few people knew something was wrong, but they were in a small minority and really didn't know what exactly was wrong or how to demonstrate the problem. As Kent Gillingham mentioned in the quote found at the beginning of this story, our inability to maintain orientation in flight should not have surprised anyone . . . but it did. It surprised almost everyone. How the system changed is what the history of instrument flight is all about. Let's now look at the person who discovered a way to make the system change.

Bill Ocker

To explain the change and how it happened, we must go back to the beginning of the U.S. Air Force (called the U.S. Army Signal Corps, among other names). Figure 1.7 is a famous panoramic photo taken around 1914 from North Island, San Diego, California. It shows the entire USAF at that time. The officers are lined up in the front row, and the enlisted lined up in the back row, with the airplanes behind them. The fifth officer from the right of the photo (in the front row) is Lt. Benjamin Foulois, who on March 2, 1910, at Fort Sam Houston, San Antonio, Texas, flew the first military owned airplane as a solo pilot. His story is worth mentioning.

Lt. Foulois was given the first airplane purchased by the War Department from the Wright Brothers, a maintenance budget, and flight orders. The airplane was purchased for the sum of $25,000. The military orders were pretty simple—he was to teach himself to fly. After he and his aircrew assembled the airplane, he began to fly it on the morning of March 2, 1910. He flew four sorties that day and made a pretty hard landing on the last flight. Soon he would start writing for more flight instructions from the Wrights. Perhaps this was the beginning of what would eventually be called military correspondence courses? He also installed what might be called the first airplane seat belt. Foulois would go on to be a famous Army Air Corps General Officer, even becoming the Chief of the Air Corps. And because of his flight accomplishment, many still consider him the Father of military aviation.

(NOTE: It is rumored that there was a ten-year-old boy among the onlookers of that first military flight on March 2. The youth's name was Carl J. Crane. Carl happened to live in San Antonio and was interested in learning more about flight. He couldn't wait to be old enough to fly.)

Now you must look at the left side of the panoramic photo. Figure 1.8 is a close up of those people in the photo. The person in the middle of the photo, the tall lanky person with two medals on his chest, is Sgt. Bill Ocker. Bill Ocker would soon earn his pilot wings (license number 293) at the Curtiss

Figure 1.7 Photo of U.S. Army Signal Corps, circa 1914.

Source: Image courtesy the National Museum of the USAF®.

Figure 1.8 Enlisted personnel of U.S. Army Signal Corps, circa 1914.
Source: Image courtesy the National Museum of the USAF®.

School of Aviation and become the second enlisted pilot to do so. The first enlisted pilot to do so was Corporal Vernon Burge (Arbon, 1992, p. 22–25).

Bill Ocker is the person who deserves the credit for changing the way airplanes would be flown. Granted, it was a slow change, probably much slower than Bill would have preferred, but nonetheless he initiated the change. The story of the history of instrument flight is about the accomplishments of Bill Ocker; and I believe Carl Crane wanted us to know of the challenges that confronted Bill. Embedded within Ocker's story of instrument flight is the most significant aviation discovery of the twentieth century. Here's how it happened.

Sgt. Bill Ocker would soon earn a commission and be at the controls of military airplanes for the Army Signal Corps commanded by General John J. Pershing. During Bill's career, he would also fly with General Billy Mitchell. Bill Ocker had entered the Army as a Calvary man. He was interested in flight, and with the advent of the airplane, he wanted to get into the Signal Corps as soon as possible. He requested a transfer from his commanding officer—then Capt. Billy Mitchell. Eventually, Capt. Mitchell would also join the Signal Corps and often fly with Ocker, as attested in Ocker's flight logs. Carl Crane would credit Bill Ocker for convincing Billy Mitchell to get involved with aviation.

Another notable person in Ocker's life was a businessman by the name of Elmer Sperry. Sperry had a gyroscope company which made instruments for ships. He was also beginning to make instruments for airplanes. One type of instrument that was particularly good for ship navigation, and seemed to have application for airplanes, was the Bank and Turn Indicator. For example, when a ship's Capt. couldn't see a change of heading due to fog or darkness, the Bank and Turn Indicator provided needed information. It even had a level indicator (ball in a liquid filled tube) that displayed tilting from one side or the other. Apparently, Sperry had given one of the devices to Bill Ocker, primarily to see if it could be used during aircraft flight. The year was 1918.

Bill modified the device in order to integrate it with the airplane. He attached a C-clamp to the Turn Indicator frame, shown in Figure 1.9, which allowed him to fasten it to the strut of an airplane wing. He then added a Venturi Tube, mounted underneath the casing. The casing housed the gyro of the Bank and Turn Indicator. A hole was made in the top of the casing. When in flight, the airflow around the Venturi Tube would cause air to be pulled through the hole in the top of the casing. The air would flow over the gyro and cause it to spin. Without the use of electricity, Bill figured out a way to spin the gyro of the Bank and Turn Indicator.

Figure 1.9 Modified Turn Indicator.

Source: Image courtesy the AF Material Command History Office.

Bill would carry this indicator with him most of the time he flew. He kept it in a little black Dopp kit. When he would go to the airplane, he'd clamp the instrument on a vertical strut near the cockpit and go fly. He would record his experiences. He reported on several occasions that the indicator seemed to work well in clear weather, but it often acted up when flying in clouds. On more than one occasion, Bill sent the device back to Sperry to have the gyro checked. It always came back, bench checked fine . . . fly as is. Bill continued to fly with this device for several years, and he would continue to document its performance, never really sure if it had any value or if it was even giving correct information. Regardless, he continued to use the instrument for approximately eight years before making a remarkable discovery. The discovery not only changed his life but it would change the lives of several others, and also change the way we would fly airplanes. The discovery was enormous.

David Myers

While assigned to Crissy Field, California, Bill Ocker continued to fly with his Bank and Turn Indicator, often frustrated with its performance. He never realized the stage was being set for an awakening. The awakening began during one of his semi-annual physicals in 1926, which included a rotating chair test that utilized a device known as the Barany Chair. The administering flight surgeon, Capt. (Dr.) David Myers (see Figure 1.10), upon completion of the routine physical, asked if he could give Bill another test (Ocker & Crane, 1932, p. 13–14).

Figure 1.10 Capt. (Dr.) David Myers.

Source: Image courtesy the AF Material Command History Office.

Earlier Dr. Myers had made an interesting observation. After spinning pilots in the Barany Chair to check the functionality of their inner ear, a routine part of the physical exam which insured the pilot had a healthy vestibular system, he noticed that if he spun them with their eyes closed for a short while, they couldn't tell which way they were spinning. He noticed further that when pilots slowed down from a constant rotation, and kept their eyes closed, they seemed to get confused as to which way they were rotating. Most reported feeling they were turning in the opposite direction. This was not part of the required physical examination, but it was something of interest to Dr. Myers. And he would routinely ask pilots to volunteer to participate in his test.

Since Bill had already passed his physical with Dr. Myers, he had no problem participating in the doctor's non-threatening, post-physical examination. Dr. Myers asked Bill to wear a blindfold while being turned in the Barany Chair. Bill went along with it, but unknown to Bill, soon after rotating, Dr. Myers slowed down the chair's rotation and asked Bill which way he was turning. Bill responded to the direction he felt he was turning. And once he indicated the direction, Dr. Myers removed the blindfold. Much to Bill's surprise he had indicated exactly opposite to the direction he was turning!

Bill had trouble believing what he just experienced. What triggered him to request another test is not certain. Perhaps it was something to do with his flight experience using the Sperry Bank and Turn Indicator, or maybe it was just a stroke of genius? Regardless, Bill asked Dr. Myers if he could return and take the test at another time. He needed to think about what had just happened. Dr. Myers said to return anytime. He enjoyed showing pilots they didn't know their left from their right.

Soon after departing Dr. Myers' office, Bill devised a contraption that would allow him to test his sensation against reality. He rigged up a wooden box about the size of a large shoe box. The box was opened on one end so he could put his head in the box and look toward the opposite side of the inside of the box, where he mounted the Bank and Turn Indicator. He taped a flashlight in the box so he could see the direction of the Turn Indicator. He finally connected a tube to the back of the Bank and Turn Indicator so he could blow on the gyro to make it spin, similar to how he figured out a way to spin the gyro with the airplane. Once he had put it all together, he returned to Dr. Myers to take the test again. But this time, instead of using a blindfold, he asked if it would be alright to put his head in the black box. Dr. Myers didn't mind as long as he couldn't see what was going on outside the box.

Once all was acceptable by both Ocker and Myers, the test began. Dr. Myers slowly spun Bill in the chair and at the appropriate time he asked Bill which way he was turning. Much to Dr. Myer's surprise, Bill indicated the correct direction. He did the test again, slowing the rotation before asking the direction, and asked Bill the direction he was turning. Again, Bill responded with the correct answer. And they did it again, with the same correct response. No one had been able to do this before! At this point, Dr. Myers wanted to know what Bill had in the box!

That encounter was proof positive of Macready's claim—you needed something outside your sense of direction to correctly determine one's orientation. You could NOT do it without an external aide or some kind of flight instrument. This was the first ever discovery of what would eventually become known as spatial disorientation—a conflict between a pilot's perception of aircraft motion and reality. It was a physiological event no one had ever demonstrated. Bill was now convinced all the techniques that many of his pilot peers had developed to help them fly in clouds and at night, when they couldn't see the horizon—shaking their heads, holding the stick between your legs, etc.—were wrong. He now had proof positive, and he took it upon himself to convince others of the problem. It would become a most challenging journey too, one he would continue throughout his military career.

Allow me to elaborate just a bit. The perceptual phenomenon may have been noticed by others at some time, but no one had ever correlated this phenomenon with the problem pilots were experiencing in flight. And full credit should be given to Dr. Myers for developing the test which triggered the idea. (Dr. Myers often felt he was never given adequate credit for this novel discovery.) The Medical Corps were doing their best to insure only the best pilots would fly. Of course, the selection process wasn't helped when pilots failed to share the experiences they were having when flying in these obscure visibility conditions, but that's the way it was. Bill had stumbled onto something that would impact everyone, and he devised an instrument to demonstrate the problem in a safe manner.

There was no longer a need to guess that something was wrong; Bill had proof you couldn't do it by the seat of your pants. And it wasn't because you weren't made of the right kind of stuff. No one could do it, even though many still felt they could. For the first time, there was definitive proof that ANYONE, pilot or not, could be fooled as to their direction of turn, if they relied upon their sense of direction. Even today it's difficult to grasp that a person could actually feel as if they are turning one way, yet in fact they are turning in the opposite direction, but that's the way it works. And because his discovery was contrary to common thinking, Bill would experience many unjustified attacks on his character and sanity.

The contraption Bill constructed was pretty crude (called the Vertigo Stopper Box or Ocker Box, Figure 1.11) even by the standards of the day, but it was effective. He worked on making improvements and eventually filed a patent. The patent for this instrument was awarded to Bill in February 1928 by the U.S. Patent Office.

One could now pass the Dr. Myer's rotation test, using the Vertigo Stopper Box. When asked which direction you're rotating in, you had to respond with the direction of turn as indicated by deflection of the turn needle and NOT by what one was feeling. Simply put, you had to overcome the sensation of turning by relying on the "flight instrument." Sounds simple, but overcoming those sensations can be difficult, even extremely difficult in some situations. Aviation was just about to take the next big step in its evolution—it was now

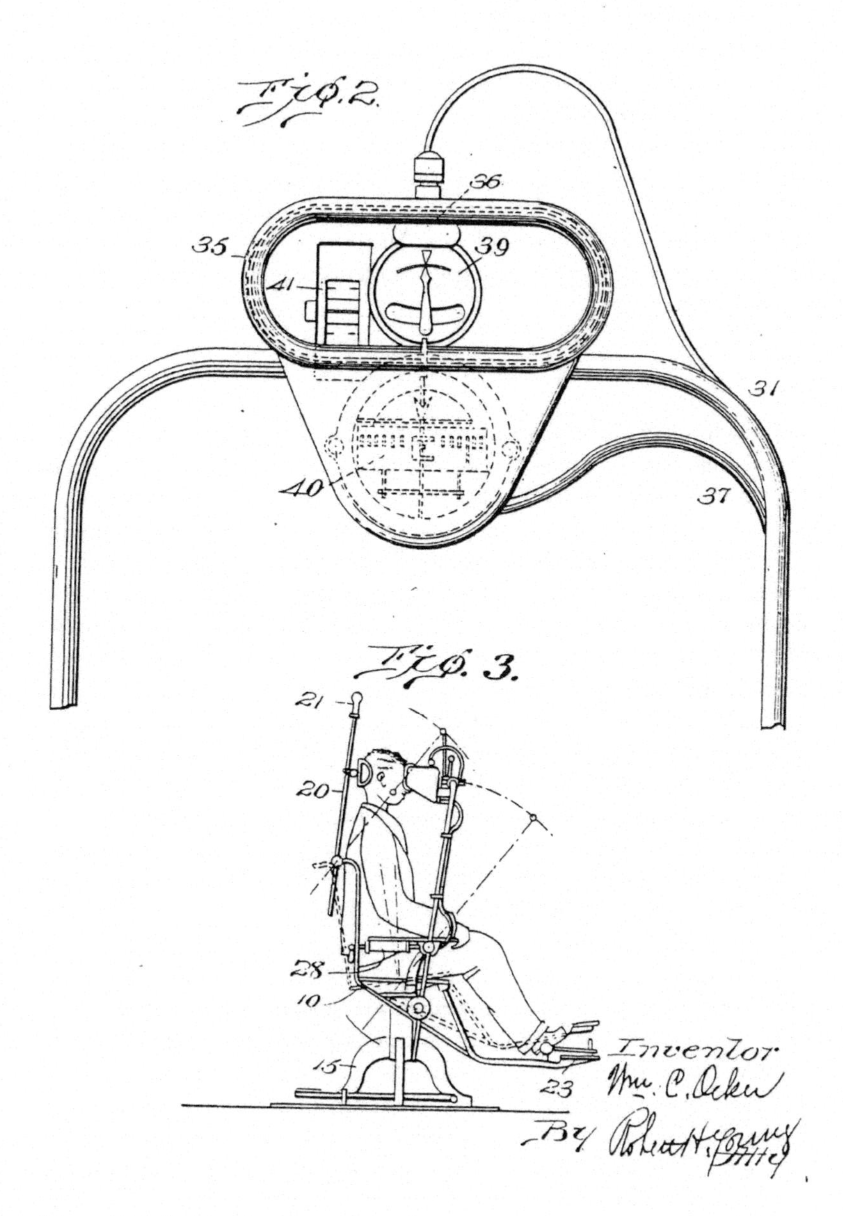

Figure 1.11 Patent of vertigo stopper box.

Source: Image courtesy Carl Crane.

possible to safely fly an airplane in just about any visual conditions, except in violent thunderstorms, as long as you had a flight instrument to rely on for spatial orientation.

As explained earlier, not many pilots really wanted to admit they needed an instrument to fly an airplane. Bill had to bear the brunt of much ridicule by his peers, and even suffer accusations by his commanders that he was crazy. Bill was sent to Letterman General Hospital for psychiatric evaluation, which he passed. (And to add to Bill's humiliation, he would be sent a second time for another psychiatric examination. Again, he passed.) Regardless, Bill continued his advocacy of the need for flight instruments in order to safely fly airplanes.

Perhaps his best piece of luck was about to happen when he was sent to Brooks Field (a flight training base), if for no other reason than to get him away from the operational pilots. While at Brooks, he happened to meet Lt. Carl Crane, and the real work of using instruments for safe flight would take another giant step forward.

Carl Crane

Upon flight school graduation from Kelly Field and Brooks Field, Lt. Carl Crane was assigned to Selfridge Field, Michigan. During his tour of duty, he was given the assignment to fly a congressman's son from Selfridge Field, Michigan to Bolling Field in Washington, DC. In the early morning hours, he picked up his passenger at base operations and proceeded to the airplane. Carl would fly in the front seat and his young passenger would fly in the rear seat. The aircraft was a JN-4, known by the nickname Jenny. The preflight, taxi, and takeoff were uneventful. During the climb-out, Carl saw some fog in the distance and he thought it would not be a problem to climb above it. As he approached the fog, the rate of climb of the airplane decreased until it wouldn't climb anymore. He maxed out around 7,000 feet, and he was not able to clear the top of the fog/clouds. He had never had any training associated with cloud flying, and he never thought much about it. But here he was at 7,000 feet and the airplane wouldn't climb any higher. He had a decision to make—turn around quickly to avoid the clouds, or enter the clouds and wait until the airplane flew through it. He chose the latter.

Not too long after the Jenny entered the fog, and while Carl believed he was holding a steady heading and altitude, he noticed something odd. The altimeter began to decrease and the airspeed began to increase. There was no attitude or Bank and Turn Indicator in the airplane—the attitude indicator had not yet been invented since there was no known need for one. And you know the reason there was no Bank and Turn Indicator—it wouldn't be used anyway. Carl then did what every pilot would do in a similar situation; he pulled back on the stick to see if this would arrest his descending airplane. Unfortunately, the altimeter continued to decrease and the airspeed indicator showed an increase in airspeed. It was getting darker and wetter (and wind rush was getting louder), and Carl was beginning to worry. He looked back at his passenger who was sitting in the rear seat. The youth was waving and having a good time. He was happy, but Carl knew something was wrong. He had no idea what it was—neither did his passenger.

Sometime during this descent, Carl tried to recall anything his flight instructors told him while in pilot training. He recalled one instructor who told him to avoid cloud flying because you never know who might be flying in the clouds, and you wouldn't want to risk the chance of inadvertently hitting another airplane. Carl should have avoided the clouds but there he was, inside the clouds and in a descending turn that would eventually be nicknamed the Graveyard Spiral (for obvious reasons), with no help from his flight training program.

Now realizing he was out of any instrument training instruction, he knew he was in serious trouble. Carl went for outside help—he prayed. That didn't resolve the situation either. The airplane continued to descend and the airspeed was increasing. Carl knew he was now dealing with the inevitable—he was going to crash! They had parachutes but no training on using them, so the idea of bailing out never crossed his mind. He decided to look back at his passenger for what he thought would be the last time. As he turned his head toward the rear of the airplane he saw a large sign go by the wing of the airplane that read "Statler Hotel." The sign was located on top of the Statler Hotel and he had just missed hitting the top of the hotel.

Now being below the fog/clouds, he was able to see the ground. He quickly leveled his wings, recovered from the descent, and was able to fly level just above the Detroit River toward Toledo, Ohio. He decided he would not fly in clouds after that experience. Something was not right with his training, and he didn't want to experience that again. The experience and the mystery would bother him for quite some time.

Note: As Carl explained it, this Graveyard Spiral was experienced by many of the early flyers. Unfortunately, they would not share their experiences because they thought it might indicate a weakness in their flying skills. Carl would eventually understand what happened to him on that flight, but it wouldn't happen until he returned to Brooks Field about a year later. There at Brooks Field as an instructor pilot he would meet and come to appreciate the work being accomplished by Bill Ocker.

Bill Ocker and Carl Crane

At Brooks Field, Bill Ocker was in the midst of a battle. Many pilots thought his campaign for better instruments was unnecessary. Bill knew instructor pilots were teaching "seat-of-the-pants" flying, and he constantly struggled to correct the situation. The photo shown in Figure 1.12 was taken by Bill shortly after he arrived at Brooks Field. The flight instructors had pasted paper over the most important instrument in the cockpit at that time—the Bank and Turn Indicator. Instructors, not aware of Bill's discovery, wanted their students to develop seat-of-the-pants flight proficiency, and to do so it required the development of their natural abilities, not the use of flight instruments. In spite of the ridicule, Bill would give demonstrations with his turning chair and box whenever the opportunity presented itself. Fortunately, the flight surgeons from the USAF School of Aviation Medicine, also located at Brooks Field at this time, were coming to appreciate Bill's Vertigo Stopper Box. They seemed to be more interested in the contraption than the pilots.

Note: Carl mentioned that Bill even gave a demonstration to Charles Lindberg once while at Brooks Field. Charles agreed to take the demonstration in the Barany Chair, but it had to be done in a tent with no one watching. At the completion of the demonstration, Charles thanked Bill for the demonstration, and for taking time to show him the Vertigo Stopper Box. Nothing more was ever said about it.

Bill remained persistent to this calling for better flight instruments. It may have been a burden for him, but it was a burden he willfully accepted. As fate would have it, Bill happened to be demonstrating one of his turning chair tests to another pilot just as Lt. Carl Crane walked into the Hangar. Carl watched the demonstration and then after observing the outcome, he decided to ask Bill to explain what was going on. Bill gave Carl the demonstration, and then he showed him how using the flight instrument could help overcome the misperception. You had to make yourself believe the instrument and not your feelings. Carl came out of that demonstration a changed person. Bill had just

Figure 1.12 Photo of airplane instrument panel, Brooks Field, TX, circa 1928.
Source: Image courtesy the AF Material Command History Office.

Figure 1.13 Carl Crane (front seat) and Bill Ocker (rear seat).
Source: Image courtesy the AF Material Command History Office.

explained to Carl what almost killed him and his young passenger when Carl flew into the fog while departing from Selfridge Field. The airplane happened to start a turn and Carl could not correctly sense he was turning and he didn't have instruments to help him. Carl now knew what had to be done, and the two became life-long friends.

They conducted some very remarkable flights under the hood (a vision restricting device that could be pulled up over the cockpit so the pilot could not see outside). And together they would advocate the use of flight instruments while conducting numerous demonstrations that proved to the pilot the need for flight instruments. In 1930, they would make the first ever, instrument only (i.e., hooded), cross-country flight from Books Field, Texas to Scott Field, Illinois (see Figure 1.13).

These two pilots could not have found each other at a more appropriate place and time. Brooks Field was not only the place for all military pilots to learn their flying skills, it was home for the School of Aviation Medicine (SAM). It was a perfect mix—instructor pilots, flight surgeons, and student pilots, all located in one place. Bill and Carl would continue to advocate the use of flight instruments, and with the help of flight surgeons and eager students, they would conduct numerous demonstrations to prove to pilots the need for using flight instruments when visibility was marginal or obscured. Many of the pilots thought it was a waste of time, but some would listen carefully. Bill and Carl devised some pretty thought-provoking tests to prove their theory.

The pigeon story

Carl happened to have a friend who flew airplanes out of Stinsons Field. Stinsons Field was less than a couple miles from Brooks Field. His friend always claimed that he could fly like the birds, which didn't need instruments to fly. He was smarter than birds—at least that was his thinking. Most pilots felt this way, so there was some logic for his attitude. Carl got an idea. The idea may have come from him observing ducks on foggy mornings. The ducks always seemed reluctant to fly into the fog. They usually stayed on the ground. So he and Bill borrowed some homing pigeons from Fort Sam Houston to test the idea. They would hoodwink the pigeons by placing a tobacco sack over their head (see Figure 1.14). They would then take the pigeons up in the airplane, toss them out of the cockpit, and observe their flight. Carl said the pigeons would start to flap their wings soon after being tossed out of the cockpit, but would stop flapping after a few attempts, and they would extend their wings in a high dihedral angle. The pigeons would hold their wings in this dihedral and descend somewhat controllably down to the ground. The wing position acted as a parachute. Once on the ground and the hood was removed, the pigeons would immediately fly back to Fort Sam Houston. Bill and Carl had

Figure 1.14 Hooded pigeon.

Source: Image courtesy the AF Material Command History Office.

just demonstrated that even birds needed something to see in order to fly safely. Without vision, birds do not fly any better than man.

The straight line story

As has been said several times previously, it was a common understanding that most pilots did not need instruments to fly an airplane. The really good pilots had that innate ability to fly; at least that was the thinking at the time. To show it was virtually impossible to stay oriented without the use of some visual reference, Bill and Carl devised a simple demonstration that is still in use today. They would recruit volunteer pilots from the flying squadrons at Brooks Field and ask them to participate in a blindfolded walking test (see Figure 1.15). The subjects would be blindfolded and asked to push a chalk liner (a device used to mark off the boundaries of a baseball field) in a straight line across the open flight field.

Their task would be to concentrate on their direction and walk as best they could in a straight line across the field. The chalk liner would dispense a small line of chalk as they walked. To document the actual outcome of the test, Carl and Bill would get in an airplane and take airborne photos of these "straight lines" (see Figure 1.16). No one could walk blindfolded in a straight

Figure 1.15 Two volunteers ready to walk a straight line.

Source: Image courtesy the AF Material Command History Office.

Figure 1.16 The resultant "straight" line.
Source: Image courtesy the AF Material Command History Office.

line. Some pilots would spiral left and others right, but everyone would spiral. Little by little, it was becoming more apparent that pilots needed instruments for orientation.

The book

After accumulating all this knowledge, Bill and Carl decided to put together a book that explained a lot of what they learned and believed to be true. The instrument concept was becoming a little more acceptable, and there were others who were beginning to recognize the importance of the discovery. However, no one had assembled in one place a book that covered such topics as blind flight problems and training procedures, instruments for spatial orientation, avigation (*sic*) aids and instruments, and blind fight training and operations. The book was first published by the Naylor Publishing Company of San Antonio, Texas in 1932. A photo of the table of contents is shown in Figure 1.17.

The book was an immediate success outside the military. It was translated into several languages and used throughout the world as the foundation to build some of the first instrument training programs. Bill even helped the blossoming commercial airline industry by consulting with the Mexican Division of Pan American Airways at Brownsville, Texas. Pan Am was starting up their

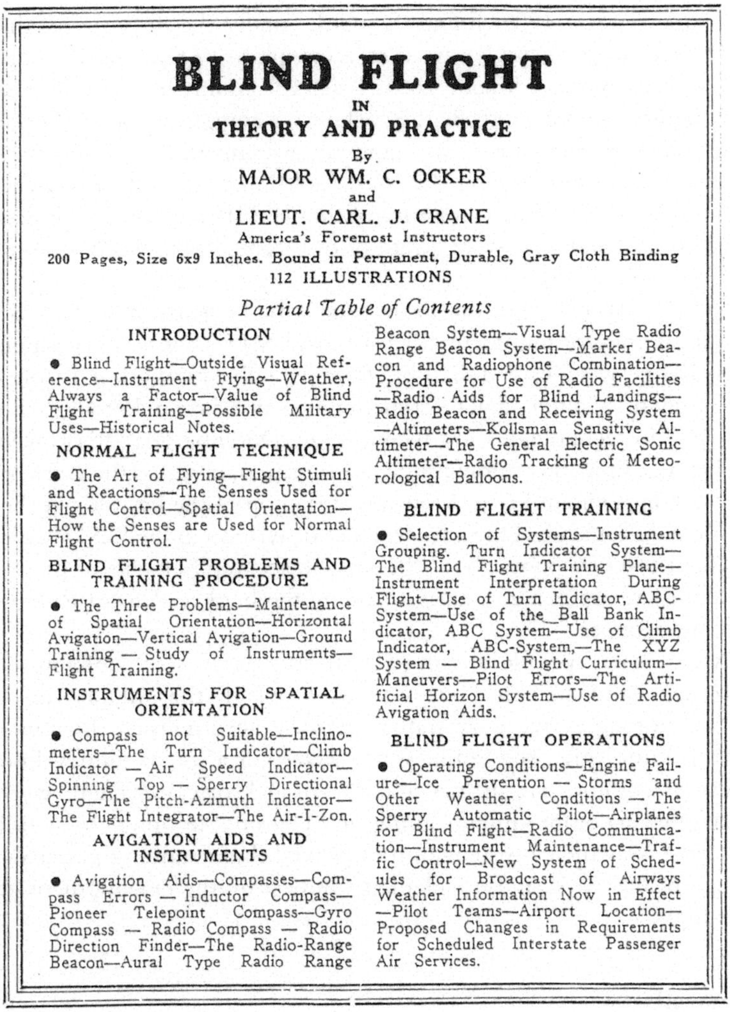

BLIND FLIGHT
IN
THEORY AND PRACTICE
By
MAJOR WM. C. OCKER
and
LIEUT. CARL. J. CRANE
America's Foremost Instructors
200 Pages, Size 6x9 Inches. Bound in Permanent, Durable, Gray Cloth Binding
112 ILLUSTRATIONS

Partial Table of Contents

INTRODUCTION

• Blind Flight—Outside Visual Reference—Instrument Flying—Weather, Always a Factor—Value of Blind Flight Training—Possible Military Uses—Historical Notes.

NORMAL FLIGHT TECHNIQUE

• The Art of Flying—Flight Stimuli and Reactions—The Senses Used for Flight Control—Spatial Orientation—How the Senses are Used for Normal Flight Control.

BLIND FLIGHT PROBLEMS AND TRAINING PROCEDURE

• The Three Problems—Maintenance of Spatial Orientation—Horizontal Avigation—Vertical Avigation—Ground Training — Study of Instruments—Flight Training.

INSTRUMENTS FOR SPATIAL ORIENTATION

• Compass not Suitable—Inclinometers—The Turn Indicator—Climb Indicator — Air Speed Indicator—Spinning Top — Sperry Directional Gyro—The Pitch-Azimuth Indicator—The Flight Integrator—The Air-I-Zon.

AVIGATION AIDS AND INSTRUMENTS

• Avigation Aids—Compasses—Compass Errors — Inductor Compass—Pioneer Telepoint Compass—Gyro Compass — Radio Compass — Radio Direction Finder—The Radio-Range Beacon—Aural Type Radio Range Beacon System—Visual Type Radio Range Beacon System—Marker Beacon and Radiophone Combination—Procedure for Use of Radio Facilities—Radio Aids for Blind Landings—Radio Beacon and Receiving System—Altimeters—Kollsman Sensitive Altimeter—The General Electric Sonic Altimeter—Radio Tracking of Meteorological Balloons.

BLIND FLIGHT TRAINING

• Selection of Systems—Instrument Grouping. Turn Indicator System—The Blind Flight Training Plane—Instrument Interpretation During Flight—Use of Turn Indicator, ABC-System—Use of the Ball Bank Indicator, ABC System—Use of Climb Indicator, ABC-System,—The XYZ System — Blind Flight Curriculum—Maneuvers—Pilot Errors—The Artificial Horizon System—Use of Radio Avigation Aids.

BLIND FLIGHT OPERATIONS

• Operating Conditions—Engine Failure—Ice Prevention — Storms and Other Weather Conditions — The Sperry Automatic Pilot—Airplanes for Blind Flight—Radio Communication—Instrument Maintenance—Traffic Control—New System of Schedules for Broadcast of Airways Weather Information Now in Effect—Pilot Teams—Airport Location—Proposed Changes in Requirements for Scheduled Interstate Passenger Air Services.

Figure 1.17 Blind flight table of contents.

Source: Image courtesy the AF Material Command History Office.

commercial flights to Mexico, and it is believed that Pan Am was the first school to adopt Ocker's blind flight "hooded" training program.

Carl learned later in his career that the Russians also had a copy of the book and had it translated for use in their flight instrument training programs. This textbook appears to have been used by almost everyone who was interested in better flight training with one possible exception—the U.S. Army Air Corps. The Army was reluctant to change. Change would eventually take place, but it would take the persistence of Bill and Carl, and time, for it to happen.

Jimmy Doolittle

About this same time in 1929, Jimmy Doolittle, under the direction of the Daniel and Harry Guggenheim Foundation, accomplished a remarkable feat at Mitchel Field, New York, (Hoppes, 2005, 101–113). Jimmy managed to take off and land without looking outside the cockpit . . . completely blinded from any outside visual information. It was quite an accomplishment. The exact level of shared information among Bill, Carl, and Jimmy is not well known, but Carl did say that Doolittle was aware of the work being done at Brooks Field. Likewise, Bill and Carl were aware of the accomplishments by Jimmy Doolittle.

Due to Jimmy's success with this project, he would eventually receive the first aeronautical engineering doctoral award from MIT. It was not without controversy, but he was awarded the PhD from MIT for this work in accomplishing the truly first "blind" flight—something we don't even do today with all the advances in technology! And when Jimmy Doolittle was interviewed later in his life regarding his accomplishments in aviation, he stated the work he did under the Guggenheim Foundation was at the top of the list. He felt this "blind" sortie did a lot toward the advancement of aviation in the U.S. Should the reader want to know more about the specific accomplishments of Jimmy Doolittle, and there are many, I suggest reading the book *Calculated Risk* by Jonna Doolittle Hoppes.

There were other people involved with the early development of flight instruments and their use, but most were outside the military. Even Doolittle was not on active duty status when he accomplished his blind sortie. At this same time, the country was investing a lot of money into the business of commercial airlines and the model airways. There were many remarkable accomplishments going on. The unique thing about Bill and Carl's work was their insistence on improved flight training practices. This was significant work, and it had a large impact on the development of instrument flight.

James Burwell

In early 1930, the Air Corps Training Center released a statement to counter a lot of the effort Ocker had been advocating. The official position for instrument training is summarized by the statement which reads:

> Blind flying is not necessary under normal conditions, and is extremely dangerous under abnormal conditions with the instruments we now have. It is strongly recommended that blind flying be not included in any phase of training at the Air Corps Training Center.
>
> (Purificato, 2012, Episode 3)

This doctrine must have been frustrating to both Bill and Carl, but it didn't seem to bother them. They continued to advocate the need for better instruments and training practices throughout the Air Corps Primary Flying School.

Fortunately, for many of the soon-to-be pilots, and in spite of the directive, the Commander of Air Corps Training Center, General Frank Lahm, eventually listened to what Bill and Carl were preaching, and by the end of the year, he ordered the graduating class held over to receive four hours of "bag" training. Although these pilots had already received their wings, they could not advance to their first flying assignment without another four hours of instrument training. It must have been a frustrating ordeal for the pilots. They had just graduated from flight school and were eager to get on with their careers, but now, before moving on to their first assignment, they were required to complete Ocker's and Crane's instrument training program. This marked the beginning of a formal flight instrument training curriculum for military pilots.

Not all pilots ignored the need for instrument training. One in particular was a young 2nd Lt. James "Jimmy" Burwell. Jimmy had almost killed himself when trying to fly in some fog. He fortunately recovered the airplane just before ground impact and was able to make a somewhat hard landing. The airplane was not damaged and he was able to get out of it. Once egressed from the airplane, he got on the ground and kissed it. It was something he would remember for the rest of his life, and a vivid story he would share whenever asked about the early days of learning to fly.

Jimmy's encounter with fog happened on a flight very similar to what happened to Carl Crane when he was flying a congressman's son from Selfridge Field to Bolling Field. However, Jimmy was solo. He was spit out of a cloud bank in some kind of Graveyard Spiral, not knowing what was going on, and fortunately he was able to see the ground and immediately recovered to make a hard landing in the field below. He chose not to tell anyone about the incident until many years later. The odds are pretty good many other pilots must have had similar experiences. When Lt. Burwell received the additional flight time under the hood, he took the training seriously.

The certificate photo in Figure 1.18 is Jimmy's graduation certificate. These certificates were given to all pilots who received the additional four hours of flight training. As you can tell by the artwork on the certificate, most still felt the training was more of a joke than something of value. They only completed the training because they had to do it to get their wings, not because they wanted to do it. The photo in Figure 1.19, of then Major General (retired) Jimmy Burwell, was taken at Air Force Village II around 1998. He was in his late nineties at the time. He was still very proud of that certificate, and he wanted others to know about his story and the way we used to do instrument flight. He was very thankful for the work of Bill and Carl.

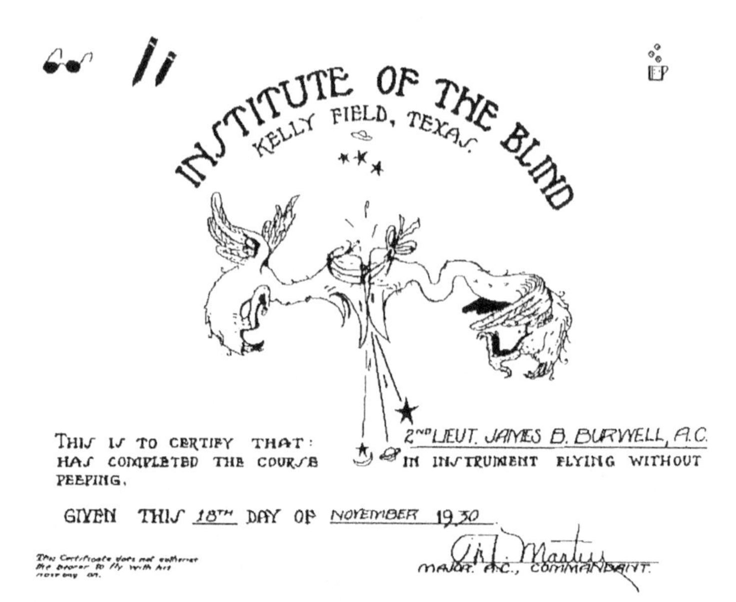

INSTITUTE OF THE BLIND
KELLY FIELD, TEXAS.

THIS IS TO CERTIFY THAT: 2ND LIEUT. JAMES B. BURWELL, A.C.
HAS COMPLETED THE COURSE IN INSTRUMENT FLYING WITHOUT PEEPING.

GIVEN THIS 18TH DAY OF NOVEMBER 1930

MAJOR. A.C., COMMANDANT.

Figure 1.18 Certificate of completion for blind flight training.
Source: Image courtesy Jimmy Burwell.

Figure 1.19 Photo of Jimmy Burwell.
Source: Personal photo.

The big three

There are many photos of Bill Ocker, Carl Crane, and/or David Myers, but it's difficult to find one with all three of them together. I've been able to find only one, and I'm not absolutely certain it's the three of them, but it appears to be. Why this photo was taken is unknown . . . perhaps posing for a news release? They were starting to receive notoriety in the local San Antonio area for their accomplishments. The photo in Figure 1.20 contains Bill Ocker (for sure) in the front, David Myers sitting on one of the Barany Chairs next to the airplane (likely), and Carl Crane (for sure) sitting in the back seat of the Douglas O-2 airplane. A hood can be seen over the front cockpit. The photo was taken at Brooks Field, perhaps close to Hangar 9, which is still in existence today (the only remaining wooden World War I hangar in its original location).

These three people and the location are so fitting for what would evolve. Brooks was then the hotbed of instrument flight training, but this would change soon. Randolph Field was under construction and many of the airplanes, those involved with the flight training mission, and the school (SAM) would move to Randolph Field in 1931. Brooks Field would take on other military roles, but the flight training mission would be gone from Brooks Field forever. Carl went to Randolph Field with the first squadron of airplanes, but Bill would not make that trip. His career continued at Brooks Field for a while and then

Figure 1.20 The big three: Ocker, Myers, Crane at Brooks Field, circa 1929.
Source: Image courtesy the AF Material Command History Office.

he was eventually assigned to Chanute Field. Airplanes would continue to fly out of Brooks until around 1960, but they were primarily in support of military airlift and research. Ironically, the USAFSAM and its mission would return to Brooks in the early 1960s to take on the role of supporting our country's effort to go to the Moon. A new complex of buildings for USAFSAM was constructed at Brooks Air Force Base, and the complex was dedicated for this new role by President John F. Kennedy on November 21, 1963.

Bill's legacy

While Bill remained at Brooks Field, he continued to push for more instrument flight training. He was not the favorite pilot of the base commander. Actually, the commander thought Bill should not fly airplanes any longer. He was older than most pilots at the time, and this instrument need was still in its infancy and not really accepted by the mainstream flight training within the military. Bill was still advocating doing things contrary to the norm.

The commander discussed his concern with the flight surgeon who would be giving Bill his physical examination, and a satisfactory examination was needed for Bill to continue to fly airplanes. The commander mentioned that one should be careful when giving older pilots their physical checkups, especially the eye examination. Rumor has it that he implied to the flight surgeon that Bill should fail the eye exam. The exact instruction is unknown, but we do know the flight surgeon failed Bill on the eye exam. Bill then made the comment to the flight surgeon that no one, including the base commander, could have passed the eye exam. The comment was relayed back to the commander, and the commander felt the comment was within the definition of conduct unbecoming an officer. He decided to put Bill up for a court martial, unless he would withdraw the statement. Bill would not withdraw his statement, claiming it was a true statement. However, he did agree to apologize for making it. This was not good enough to satisfy the commander, so they went to trial.

Fortunately for Bill, he was able to get good legal help. His lawyer contacted several of the more famous pilots who appreciated Bill's impact on their flying professions. During the trial, written testimonies were given by Orville Wright and Amelia Earhart. Both gave testimony regarding the significance Bill had on safely advancing aviation for our country and throughout the world. It has been stated that Orville Wright wrote a letter explaining that except for Maj. Ocker's great zeal as a missionary, Orville doubted whether the course in blind flying would be a requirement in the Army today. Orville believed that Bill's campaign of education has had more influence in bringing about the use of instruments than that of any other person. The trial lasted a little over a week, and Bill was ultimately acquitted.

Note: An observation is appropriate here. Billy Mitchell and Bill Ocker, two noted flying experts, had similar personalities and their military paths crossed several times. From Bill's flight log books, he was an instructor pilot for General Mitchell on several occasions. When General Mitchell was a Captain in the U.S. Calvary, he was Bill Ocker's commanding officer. Both had a vision and would not deviate from making others realize the importance of their convictions. And because of these strong convictions, they ran into trouble with the establishment. Billy and Bill each ended up suffering through an independent court martial. Ultimately, both were exonerated and their ideas have withstood the test of time. There has to be something to be learned about human nature by studying these two individuals.

Bill would serve an assignment at Chanute Field, Illinois and eventually retire in 1941. Figures 1.21a and b depict an early photo of Bill around 1913 and his last official military photo—displaying the two medals on his chest (marksmanships for pistol and rifle). They cover a span of roughly thirty years (1913 to 1943). It's remarkable what he accomplished in that short time period. When he retired, he had accumulated more flying time than anyone else on active duty at that time. It's not certain, but we can assume by his many in-flight accomplishments he most likely had more "bag" time than any of his peers. He inspired others to continue his quest for better flight instrument training. But many of these accomplishments were sandwiched around a few frustrations. During his military career, he was sent to Letterman General Hospital for a

a

b

Figure 1.21(a and b) Two photos about thirty years apart: Bill Ocker.

Source: First image courtesy the AF Material Command History Office; second image courtesy Carl Crane.

psychiatric evaluation on two separate occasions. He would eventually joke about how he was the only officer in the entire Air Corps who could actually prove he was sane—he had two letters to validate the claim. No other officer in the Air Corps could make such a statement!

Bill Ocker was born in Philadelphia, Pennsylvania on June 18, 1880 and passed away, not long after retirement, on September 15, 1942. He was sixty-two. Many ridiculed his insistence on needing to trust your flight instruments, but some did take his lessons to heart. One of his supporters and students by the name of Colonel Joe Duckworth was given permission to open a school dedicated to training pilots about the need for using flight instruments for airplane flying. The school opened at Bryan Field, Texas at the height of World War II, and it was named the Instrument Pilot Instructor School or IPIS. Graduates from the school received a Green Card which indicated they were graduates of this important instrument flight training program. Eventually, with the advent of consolidated undergraduate pilot training, instrument training was integrated with basic and advanced pilot training, eliminating the need for a separate instrument training program. IPIS, and its parent organization The Instrument Flight Center, located at Randolph AFB, Texas, officially closed in 1978.

The same Instrument Flight Center (IFC) that closed the first time in 1978 would eventually reopen at Randolph AFB in 1983 without IPIS. And not long after reopening, the new IFC eventually merged with a few other flight procedure functions and all together these functions formed what is now called the AF Flight Standards Agency (AFFSA). Within the AFFSA there exists an offshoot of the former IPIS, known as the Advanced Instrument School or AIS. The AIS is located at Tinker AFB, Oklahoma.

An excerpt from an early instrument flight training manual found in the history files of Air Education and Training Command, Randolph AFB, Texas illustrates how long it takes to change certain beliefs. It mentions that if you experience spatial disorientation (a new term to most pilots) during instrument flight, you should relax, both manually and physically, and possibly laugh at yourself. Shaking your head and moving your body around will help you eliminate the unpleasant sensation. Really?! As you may guess by now, it took a long time to remove many of the myths and ideas about how best to use instruments to fly an airplane. The only statement that has withstood the test of time is to believe in what your instruments are showing you and react appropriately.

Carl's legacy

Carl Crane was born in San Antonio in late October 1900. At the age of ten, he was able to watch Lt. Benjamin Foulois fly the Wright Military Flyer out of Fort Sam Houston. Carl was always interested in learning to fly. Whenever he could he would ride his bike south of town to Brooks Field to watch the airplanes take off and land. He couldn't wait to learn to fly! As soon as he became of age, he joined the Army Signal Corps and became a pilot. His early flying skills were learned at Kelly Field and Brooks Field.

Figure 1.22 Colonel Carl J. Crane, March Field, CA, 1947.
Source: Image courtesy the AF Material Command History Office.

After assignments at Brooks Field and Randolph Field, he was assigned to Wright Field where he eventually won a McKay Trophy in 1937 for the first fully automated landing—an accomplishment that seems even more remarkable by today's standards. He retired from active duty in 1948 after an assignment as the Commander of the First Fighter Wing, March AFB, California (Figure 1.22).

He continued to investigate flight instrument developments throughout the rest of his life. As documented in his co-authored book, *Blind Flight in Theory and Practice*, he believed the way attitude information is displayed to the pilot is contrary to human perception. He insisted that the attitude indicator was incorrectly designed. Carl continued to advocate for change until the day he died. He died in the spring of 1982 at the age of eighty-one.

Note: His concept of the correct design of the attitude indicator can be found in almost all Russian built airplanes.

One story that must be included with the history of instrument flight is the idea Carl realized while learning about instrument flight training with his

Figure 1.23 Modified Ruggles Orientator.
Source: Image courtesy the AF Material Command History Office.

colleague Bill Ocker at Brooks Field. Ruggles Orientators (see Figure 1.23) were very common at Brooks Field. These devices were intended for selecting the right type of person for flying airplanes. A pilot-trainee would be spun in the device and how quickly he could upright himself determined if he was suited to fly military airplanes. At least that was the initial thinking. After a few years of use, it screened out way too many candidates, and soon the Orientators quickly were put away. Hardly anyone could do the task correctly.

Carl got the idea the devices may be used for something else. He was given permission to modify one of the Ruggles Orientators for an instrument training idea. He added instruments inside the cockpit and put a hood over the top, much like what they did with the airplanes for instrument training. You could then control the motion of the gondola from the instructor seat shown in the photo.

He filed a patent on the idea. And soon after he filed the patent, as he explained to me, it was found to be in conflict with another similar patent. After reviewing the documents of the two inventors, the Patent Office decided Carl was two years ahead of the other person. The other person's name was Ed Link. Carl was awarded the patent for the first flight simulator. Due to certain

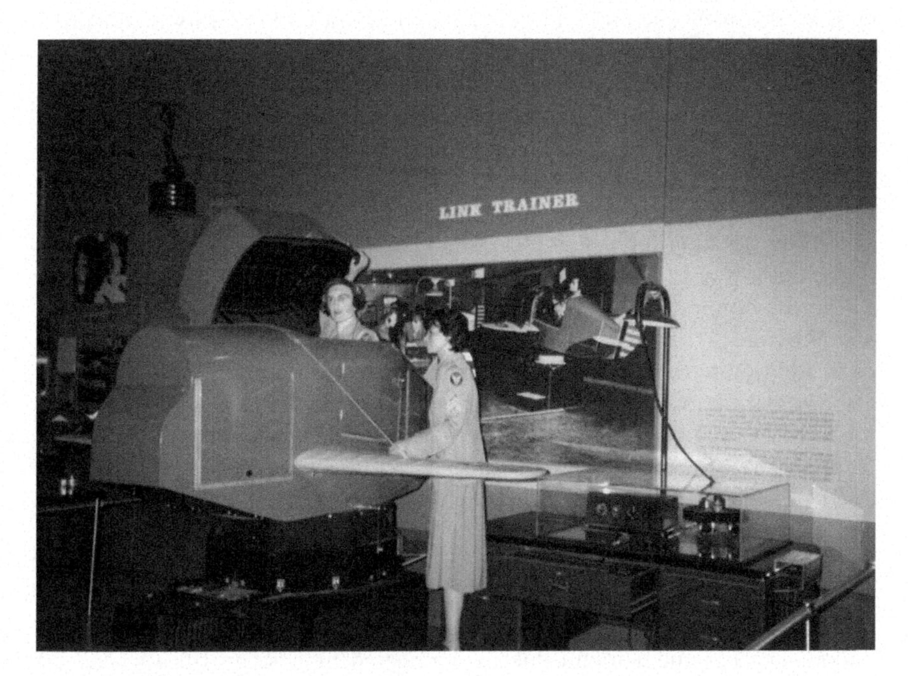

Figure 1.24 Early Link Trainer or Blue Box.
Source: Personal photo.

unclear limitations, Carl assigned his patent to Ed Link and Ed would give birth to the flight simulator industry. (Carl did say that he received royalties for the patent.) Link's new device, the Link Trainer, some called it the Blue Box (shown in Figure 1.24), would take hold and would be commonly used by just about every flight training program throughout the world. Some are still in use today! And all flight training programs today include flight simulators as an integral part of learning how to fly an airplane. Thus, the genesis of flight simulation happened at Brooks Field, Texas by Carl Crane using a device that did not do what it was intended to do (i.e., screen out pilots who did not have the right stuff for flight, Previc & Ercoline, 2004, p. 338).

Epilogue

So . . . there you have it, The history of instrument flight, about as close to what I recall how Carl told it. To the best of my knowledge, this story has never been written up completely. Personally, I was recorded several times over the years, and I know Carl was recorded at least once because I saw the video and used to have a copy. I've shared the story with many organizations all over the globe, and I often reduce the content or expand it, depending on the time available and the audience. It's quite a bit of history. There were many versions that have been told, but the basic foundation of the story is contained in the version you have in this chapter.

I would like to thank the International Society of Aviation Psychologists for giving me the opportunity to tell the story one more time, and to have it written up so others might enjoy it, or even learn from it. I always felt fortunate to have known Carl. As Pam, his wife said, Carl was a Renaissance Man. He was very creative and always looking for a better way of doing something, and always looking for a new idea. Once he got hold of an idea, he'd try to figure out a way to apply it to everyday life. He was the consummate thinker and in my opinion a true human factors expert—on the level of Paul Fitts or Stan Roscoe.

In addition to thanking and recognizing Pam, who introduced me to Doris Osborn and Jimmy Burwell, I'd like to thank Doris, Bill's daughter, for sharing her dad's photos and newspaper articles. After reading through them, the records were turned over to the Brooks AFB museum at first and then moved to Wright-Patterson AFB, Ohio, which is where they are kept today. Jimmy Burwell is due recognition because he helped elaborate on Carl's story, all the while giving insight to Bill Ocker as an instructor at Brooks Field. Jimmy Burwell was very proud of that blind flight graduation certificate. Another person who played a key role in researching the history of Bill Ocker is Lee Arbon. Lee became a friend due to our shared interests in Bill Ocker, and eventually Lee published a book dedicated to the accomplishments of all the U.S. Army enlisted pilots. The title of the book he researched and authored is *They Also Flew—The Enlisted Pilot Legacy 1912–1942*. And a special shout out goes to Dr. Kristen Liggett, a human factors researcher at the AFRL, WPAFB, Ohio, who found an original copy of the July 1924 *National Geographic* magazine, and gave me the copy. This publication contains the story of the non-stop, cross-country accomplishment of John Macready, and clearly identifies the early problem of instrument flight.

Much gratitude also goes to Ray Ortensie for his extraordinary help with finding copies of old photos, many of which were used for this chapter, and another historian, Rudy Purificato who lives and works in San Antonio. Rudy has written several very enjoyable articles about the accomplishments of many of these early aviation pioneers. He also produced an excellent video depicting the accomplishments of noted personnel who were assigned to Brooks Field, Air Force Base and/or City Base. There are many others to thank: Mike Vidulich of the AFRL and Christy Hellyer of BRKwyle for their support in getting all the information assembled in a publishable format. And a thanks to my former flight organization, The Instrument Flight Center, and the former commanders of the IFC, Colonel Jay Baker, who supported my work of learning more about this history lesson. And thanks to my family for supporting me while doing this work. Let me wrap up by telling the story of how I met Carl Crane.

I first met Carl while serving as an Instructor Pilot in the T-38 Talon aircraft at Randolph AFB, Texas around 1973/4. The exact date is uncertain. I was assigned to the newly formed USAF IFC, which was an organization made up of several divisions with different flight-related functions. My function was called the Instrument Pilot Instructor School or IPIS, as it was known among pilots. Unfortunately, IPIS and the IFC would not make it through the 1970s, being closed due to a series of post-Vietnam era budget cuts. I suspect at one time there was an important need for a separate advanced instrument training program for pilots, and IPIS served this need. But with the increased emphasis on instrument flying within undergraduate pilot training, the need for a separate school was diminishing. Post-Vietnam was here and some military flight programs were being eliminated and consolidated. Fortunately, before leaving the IFC, I met Carl.

I was teaching an academic course in IPIS called Flight Instruments. The course taught pilots about the operation and mechanics of several of the more important flight instruments. Of course, this was before the age of electronic screens that now display the same information. We were flying aircraft with what many called "round dials." As it would happen, a retired colonel by the name of Carl J. Crane, knowing the background of IPIS more than anyone could imagine, and living in nearby San Antonio, came by to audit the class (with permission from the School Commandant).

Classes took place every three weeks, so it wasn't long before Carl and I got to know each other and we began to share flying stories. And it wasn't long before he asked if he could help teach my class. I wasn't sure what that was about but I went along with it, and accepted the offer. His classroom presence, for obvious reasons, demanded your attention. Maybe it was his command of the subject? Anyway, the students loved to hear his stories, and, to be honest, I did too. He told this version of the evolution of flight instruments and how some of them came to be. He also described the human perception and how better designs could make flying safer—and easier. And he'd talk about this pilot by the name of Bill Ocker who changed our understanding of the best way to teach basic flight skills. Carl was adamant that the currently used attitude indicators were incorrectly designed and that they were the cause of several accidents. The attitude design received mixed responses from the students, but the history story completely fascinated just about everyone who listened to it. Make no mistake about it; The history of instrument flight is about Bill Ocker, but it is from Carl Crane. There was a lot more to Carl than just the history lesson, but it was the history that connected us, and through it we would become friends and fraternal brothers (Daedalians).

Note: To better understand his idea of improved attitude indicators, I accepted the offer to help test fly his attitude display (called the Flitegage). The airplane was a Cessna 172 and it contained an integrated display that was very unique and nothing like I'd ever seen before. Remember, this was 1974. Displays were mechanical and separate, not integrated. Carl had an integrated display and one that showed the bank of the airplane completely contrary to what I had learned about. After a few minutes under the bag, I was able to fly and land his airplane to 0/0 visibility conditions—without seeing anything outside!

Carl also took the time to introduce me to Brooks AFB. There he gave me a personal tour of Hangar 9. Hangar 9 remains as the only World War I wooden aircraft hangar still in existence in its original location. Brooks itself, now closed, was an auxiliary field named after Cadet Sidney Brooks, a local San Antonio citizen who was killed in an airplane accident during an attempt to land.

The base and many other bases around the country had a lot of these World War I hangars constructed during the build-up to The Great War. After World War I, and after years of neglect, the hangars could not support the newer airplanes, so they were being demolished to make room for newer buildings and facilities. Carl took exception to this action and decided to save one of the hangars for posterity. Brooks Field had a special meaning to Carl, and Hangar 9 was a symbol of those early flying days. Through a lot of trial and tribulation, Carl was able to pull off a remarkable feat by saving Hangar 9 from demolition. It was saved and restored in the late 1960s and the Hangar still stands today as a symbol echoing the names of those World War I pilots, and many others, who trained and worked out of Brooks Field. The Hangar has been closed to the public until the City figures out a way to restore it for others to enjoy. The City has recently unveiled plans for the restoration, and work is under way. Carl is happy!

In closing, Bill Ocker was never really recognized for his accomplishment. Carl always thought Bill should have received some type of recognition for his contribution to advances in aviation. It's my opinion the discovery Bill Ocker made with the rotating test devised by Dr. David Myers is the most significant aviation related scientific discovery of the twentieth century. I also believe Carl Crane and Kent Gillingham would agree with that claim.

References

Arbon, L. (1992). *They also flew, the enlisted pilot legacy 1912–1942*. Washington, DC: Smithsonian Institution Press.

Gibb, R., Ercoline, W., & Scharff, L. (2011). Spatial disorientation: Decades of pilot fatalities. *Aviation, Space, and Environmental Medicine, 82*, 717–724.

Gillingham, K. K., & Previc, F. H. (1996). Spatial orientation in flight. In R. L. Dehart (ed.), *Fundamentals of aerospace medicine* (pp. 309–397). Baltimore, MD: Williams & Wilkins.

Hoppes, J. D. (2005). *Calculated risk, the extraordinary life of Jimmy Doolittle—Aviation pioneer and World War II hero*. Santa Monica, CA: Santa Monica Press.

National Geographic Magazine. (1924). *XLVI*(1), July.

Noe, A. (1989). Medical principle and aeronautical practice: American aviation medicine to World War II (Unpublished doctoral dissertation). University of Delaware, Newark, DE.

Ocker, W. C., & Crane, C. J. (1932). *Blind flight in theory and practice*. San Antonio, TX: Naylor Printing Company.

Previc, F. H., & Ercoline, W. R. (2004). *Spatial disorientation in aviation: Progress in astronautics and aeronautics (Book 203)*. Reston, VA: American Institute of Aeronautics and Astronautics.

Purificato, R. (Producer). (2012). *The story of Brooks, episode 3*. (Matson Creative DVD), US Air Force, 311th Human Systems Wing, Brooks City-Base, TX.

2 Practitioner operational challenges and priorities

Esa M. Rantanen, Karl E. Fennell, Jason Russi, and John R. Dougherty

This chapter is based on the Practitioner Plenary Panel held at the International Symposium on Aviation Psychology (ISAP) in May 2015. The charge to the panel was for aviation practitioners to present their thoughts regarding the current and future human factor challenges in their various domains. This Plenary Panel was followed the next day by a Researcher Plenary Panel that would present research and development (R&D) approaches and strategies that address the challenges confronted by the practitioners (see Chapter 3, this volume). The overarching goal of these two plenary sessions was to foster dialogues between operational personnel and researchers in order to promote more efficient R&D efforts and a safer aviation industry.

Indeed, the foundational objective of ISAP was to foster aviation psychology work that would fit well within Pasteur's Quadrant (Stokes, 1997; see Figure 2.1). Pasteur's Quadrant was named after the French chemist and microbiologist, Louis Pasteur, who was committed both to basic research and its practical applications. This was used to illustrate that research need not be divided into distinct 'basic' and 'applied' categories. Rather, basic research should be enriched by seeking to understand phenomena as they occur in

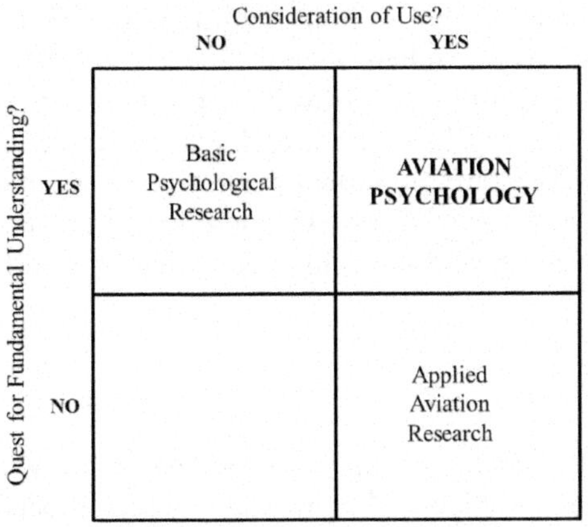

Quest for Fundamental Understanding?	Consideration of Use? NO	Consideration of Use? YES
YES	Basic Psychological Research	**AVIATION PSYCHOLOGY**
NO		Applied Aviation Research

Figure 2.1 Pasteur's Quadrant for aviation psychology.

Source: Adapted from Stokes, 1997.

the real world, and practical applications should profit from the theories and methods utilized in basic research.

As an example, consider the history of the development of instrument flight as told by Bill Ercoline in Chapter 1 of this volume. In this case, the motivation to improve the safety of flight led to a more fundamental understanding of human perception and the necessary instruments and training to overcome the limitations of human perception. Advances in both "basic" and "applied" science were made by the same people working on a specific challenge.

The panelists for the Practitioners Panel were pilots and air traffic controllers from both the civilian and military sectors. They included Richard Yasky, chief pilot in the Research Services Directorate at The National Aeronautics and Space Administration's (NASA) Langley Research Center; Jason Russi, air traffic control tower Chief Controller assigned to the 88th Operations Support Squadron at Wright-Patterson Air Force Base; Doug Glussich, Airbus A320 captain at Air Canada; Karl Fennell, Boeing B737 captain and B777 instructor and evaluator at United Airlines; and John R. Dougherty, the 119th Reconnaissance Group (RG) Commander for the 119th Wing, North Dakota Air National Guard. The panel was moderated by Esa Rantanen, an associate professor of psychology at Rochester Institute of Technology and a former air traffic controller. This chapter highlights some of the practitioners' most pressing concerns, barriers to more efficient R&D efforts to bring about a safer aviation system, and ways to move forward.

Practitioners' pressing concerns

NextGen

NextGen addresses the modernization of the National Airspace System (NAS), and many people and groups are working on new technology and solutions toward its implementation. However, the vision of how humans should interact with the envisioned new technology has not been linked up with the NextGen goals and direction. For example, although human factors researchers understand concepts like "resilience" and "cascade failures" and how automation and humans interact to cause as well as prevent catastrophes, it is the practitioners who understand how the actual systems work, problems associated with the technology, and how the operator makes things that do not always fit nicely together actually work. But there have been limited exchanges between researchers and practitioners in cooperation with industry and regulatory agencies.

Vertical Navigation functions

One challenge to the pilot in the aircraft cockpit environment is the Vertical Navigation (VNAV) function of autoflight. The onset of complex procedures such as Descend Via and Climb Via, Area Navigation (RNAV), and Required

Navigation Performance (RNP) require modern pilots to have a more sophisticated understanding of VNAV than in the past. The current VNAV is complex and difficult to use effectively. For example, to be able to fly the new complex arrival and departure procedures, pilots need to be able to identify and differentiate when the airplane is being commanded to maintain the programmed path (VNAV Path), when the airplane is above or below the path (VNAV Spd), and when the airplane is at an altitude in hold mode not on the VNAV Path (VNAV alt). And pilots would have to be doing this while developing a strategy that would return the airplane to the desired energy state if it is not in the correct mode.

The use of VNAV is especially challenging when used in areas where the margin of error can be slim. For example, using RNP procedures to fly airplanes between rugged peaks in Jackson Hole, Wyoming and to thread airplanes into complex high traffic metroplex airports in Houston, Texas (https://www.faa.gov/nextgen/ accessed November 3, 2015) is complex. Laterally, this can mean a tight turning Radius to Fix leg with airspeed limits around a peak or in a valley. The use of autoflight VNAV and lateral navigation is especially problematic for pilots who are still learning how to manage the airplane and complexities of new automation in a dynamic environment. The use of these advanced departure arrival procedures is increasing rapidly in current line operations, but both controllers and pilots may struggle with how to use this technology effectively without making critical mistakes.

Remotely piloted aircraft (RPAs): human-machine interface and automation

In recent years, there has been a trend to move the human out of the aircraft and control the aircraft from the ground. Such RPAs have been proliferating especially in military missions, such as medium and high altitude Intelligence Surveillance and Reconnaissance (ISR) missions. However, the current cockpit/Ground Control Station designs for RPA platforms leave much to be desired in the human-machine interface (HMI) arena, and are totally inadequate for any future multi-ship/swarm employment.

In addition, the role of autonomy within the RPA enterprise is just beginning to be explored. By their very nature, RPA platforms need some level of automation, and in some cases, this will approach complete autonomy. This is due to numerous reasons such as reducing operator workload, increasing precision for maneuvers such as search patterns, and dealing with communication failures, just to name a few. The need for automation drives the requirement for an HMI that is flexible and clearly communicates to the operator what the RPA is currently doing and will do in the future. As the level of automation rises, the operator will be less "in the loop" and the HMI will need to reflect this enhanced partnership appropriately. Maintaining appropriate situation awareness on the mission, communicating intent to near autonomous RPA(s), dealing with command and control directives, and compensating for fatigue, will all need to be taken into consideration. The HMI must be intuitive to use and be able to support complex flight duties that will include controlling multiple platforms in a combat environment. This reflects an important irony of advanced automation. As the operator's role shifts from being a manual controller, there is an increased need to keep the operator engaged and to support complex problem solving activities, so that the operator can intervene in those situations that were not anticipated in the design of the automated systems (e.g., fault diagnosis). Considerable research needs to be done in this arena to determine the best design strategy for such an HMI.

RPAs: sustained operations

Another RPA area ripe for immediate aviation psychology research concerns the effects of sustained RPA-ISR missions on the aircrews. The bulk of these issues stems from the sustained 24-hours-a-day/7-days-a-week/365-days-year nature of this vital combat support mission. While considerable work in the area of sustained manned airborne operations already has been done, the nature of RPA operations differs significantly from that domain and requires additional exploration. At or near the top of the priority list is the need to mitigate the long-term negative impacts associated with sustained RPA duties. The negative effects were personally experienced by one of the authors serving as a commander of an RG on a RPA-ISR mission. No previous airborne platform has introduced the same stressors and cumulative fatigue that the RPA mission has. This is due to numerous factors such as rotating schedules, poor HMI designs, combat stress, difficult weapons employment scenarios, communication barriers, inadequate SA tools, and improper manning ratios.

This reflects the need for researchers to go beyond the functional aspects of work to consider the global-user experience. We know much about how the cumulative physical stresses of work impact people (e.g., carpal tunnel). However, there is still much to learn about the cumulative stresses of the type of intense cognitive work that RPA operators are experiencing.

Rate of change and trust in novel methods and technologies

Research findings and technological innovations are often slow to be implemented in the operational environments through regulatory agencies. Operators are sometimes frustrated by not having the available technology to enhance safety, such as moving maps on Class 2 Electronic Flight Bags (EFBs). Although this product is coming to the operation, the slow churn between research, regulation, and operations could cause the operator to lose faith in the research process. In some cases, changes that were great ideas ten years ago are only now coming to pass as new methods to perform air traffic control (ATC) duties. Additionally, when new changes or methods arrive, they often are not integrated readily into the field of ATC and therefore the NAS operation. This generates significant trust issues of the new methods among controllers who are used to their traditional methods that have served them well for a long time. When new methods are introduced, the resistance to change is severe and total adoption is a lengthy process. This is not pride on the controllers' part as much as a lack of trust in new technologies and processes, especially when so much is at stake. It is unfortunate that nearly every single ATC procedural change has come about from an avoidable disaster that involved some human error. Such history can rightly induce paranoia among operators. Keep in mind that nearly all of those human factors considerations that Paul M. Fitts outlined in his 1951 paper, "Human engineering for an effective air-navigation and traffic-control system" remain as current challenges controllers face today.

Barriers to a more efficient and safer system

A number of barriers to the implementation of a more efficient and safer system are already apparent from the examples provided above. Two major classes of barriers appear to underlie the difficulties that have been presented so far: the first is the lack of human factors consideration in the design and application of new technologies or procedures, especially early in the process of change; the second is inadequate interactions between practitioners and researchers in identifying problems and devising solutions. It is shown below that a much greater extent of collaboration between practitioners and researchers than currently exists is needed throughout the process of identifying the problems, and designing and implementing corrective or enhancing measures.

Lack of human factors considerations

A basic problem with many of the evolving real-world systems is that human factors issues are often perceived to be of low priority. Consequently, responses to these issues are much delayed. Too often, the primary driver behind new systems is some disastrous consequences attributed to human error or some cost savings measures that do not take into consideration the impact on the human operator or the system that the human operator must interact with. One example is the flight management computer (FMC). The FMC was intended to reduce workload and increase efficiency by automating many tasks and utilizing advanced computations for optimized fuel efficiency. But the design of the FMC did not benefit from either insight of the research community or the operational practitioners. The result was an FMC that was complex and difficult to master for a generation of pilots. An entire new skill set had to be trained and mastered in order to take advantage of the computational power of the new FMC and to simply perform old tasks with the new system.

Similarly, with the current EFB, many applications were added without thorough user testing or adequate discussions between researchers and practitioners. Human factors standards mandated by the regulator are sometimes vague and are often applied inconsistently. New functions such as the airport moving map on the EFB, even though highly desired by the operators, were slow to be incorporated into the system because of concerns arising from a lack of complete understanding of the operators' work environment. A useable product did eventually come forward, but a more concerted effort in bringing together the practitioners and researchers early in the design phase could have significantly improved the design and effective use of the EFB much sooner.

It is fair to assume that researchers who have received their formal education from human factors-related programs in American universities are aware of the importance of user-centered design. Widely used college textbooks (e.g., Sanders & McCormick, 1993; Wickens, Lee, Liu, & Gordon-Becker, 1998) do a good job of teaching both the value and methods of user-centered design. The problem may be that there still are relatively few researchers specifically trained in human factors, that they are a minority voice in large research teams, and that the research environment lacks the means to reach out to practitioners.

Inadequate interactions between practitioners and researchers

Researchers working without adequate practitioners' inputs could have far-ranging consequences. Researchers are generally very focused and tend to perform research projects within very strict scientific parameters. It is typical for researchers to focus their attention upon a single aspect of a much larger issue. An example would be research performed on one sub-system within a much larger and more complex system, such as the NAS. An experienced practitioner can often detect a ripple effect of the way something is done currently and how that might impact many subsequent "nodes" across the larger system. Without the benefits of this foresight, the researcher might have worked long and hard scrutinizing a singular activity in developing what initially appears to be a flawless fix, only to have a practitioner immediately dismiss it due to the negative impact it will have elsewhere within the larger system. Not only would such an oversight cause a waste of time, it could significantly hurt the researcher's credibility among the practitioners. To address making a better team from the practitioner and researcher communities, it is essential to first appreciate the challenges that confront each community attempting to reach out to the other. Organizational, cultural, and language barriers are discussed below.

Organizational barriers

Practitioners and researchers each work in their own domains and have different job responsibilities that are subject to different performance appraisal and reward systems. For example, the primary job of an air traffic controller is to control traffic, just as the primary job of the commercial pilot is to fly paying passengers. Research participation will necessarily take them away from their operational duties, creating manpower issues. In the civilian arena, it is understandable that many Federal Aviation Administration (FAA) specialists charge or demand a lot of money for their time to participate in research projects.

This stands to reason because their time is valuable and they are compensated very well to perform their duties as professionals. Research participation may therefore be considered a luxury rather than a valued activity.

In addition, practitioners may be reluctant to participate in research studies as subject matter experts (SMEs) and in particular as participants in experiments. This is because of their concern of how such participation and their performance in experimental simulations might affect their relationship with their employer (e.g., the FAA) and their labor union (e.g., the National Air Traffic Controllers' Association (NATCA)). As an example, one author has had the experience of not being able to recruit air traffic controllers to participate in experimental research that was sponsored by the FAA. However, in recent years, it appears that researchers affiliated with the FAA Technical Center and NASA research centers have been more successful in having operational controllers participate in NextGen experimental simulations, perhaps signifying a trend toward closer practitioner participation in research. The published research literature does not, however, give any indication about the extent of practitioner input in experimental designs or feedback on the results.

For researchers, applied research is typically much more difficult and takes considerably longer to conduct than typical basic experimental research where college students are easily available as participants. Applied research also tends to have a lower yield in terms of publications—the primary unit of performance measurement in academia. Although there are thankfully many journals that publish applied work, they do not enjoy the same prestige in tenure and promotion decisions as more theoretically oriented scientific journals. Researchers are therefore less incentivized to take on the challenge of conducting applied research.

Further, there are structural barriers to more direct communications between practitioners and researchers. Figure 2.2 illustrates the various communication pathways among the many stakeholders in aviation psychology (researchers, research sponsors, designers and engineers, and regulators). The solid lines in Figure 2.2 represent existing, established communication pathways whereas the dashed lines represent nonexistent but desirable pathways. Figure 2.2 shows that practitioners are largely isolated from the many other players in the system. The only direct communication pathway to practitioners is from regulators in the form of regulations. Research results make their way to practitioners only through technology that has been designed by engineers and regulated by regulators. Note also that there are no solid lines leading away from practitioners to the other stakeholders. Consequently, practitioners might observe many ways to increase efficiency yet not have a readily accessible avenue to provide that input or feedback directly to researchers. At the same time, researchers desiring input from SMEs find it difficult to identify and recruit practitioners to help with their investigations. As a case in point, one author, in 15 years as a controller, has just recently been asked by a researcher how to interpret a problem and asked for input on what can be done to improve the air transportation system.

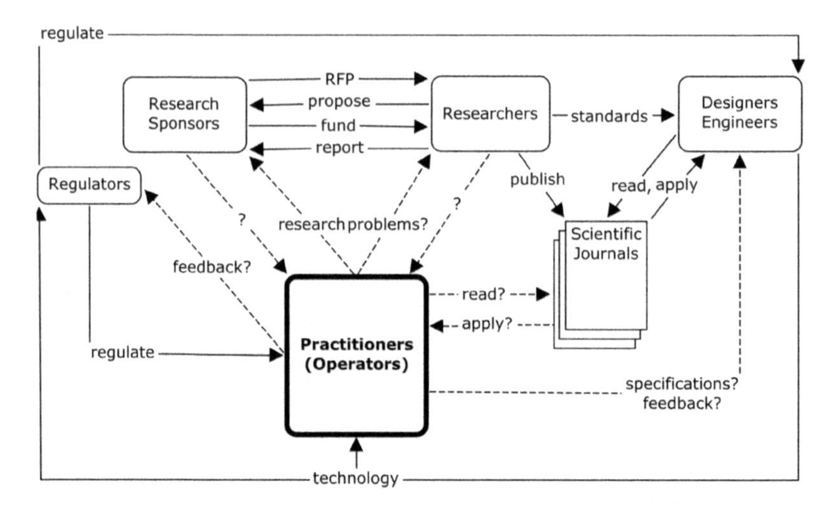

Figure 2.2 Graphical representation of communication problem between practitioners (operators) and researchers in aviation R&D.

The solid lines represent existing, established communications between different stakeholders (in rounded boxes). The dashed lines represent currently nonexistent but desirable lines of communication. Note that in this figure only direct lines to practitioners are regulations about technology, procedures and new technology, as introduced and implemented by engineers and designers. There are no communication lines extending from practitioners. The purported contributions of the plenary panels held at ISAP are depicted by the heavier dashed lines between researchers and practitioners.

There are also physical barriers. For example, ATC is a specialized field often found behind cipher-locked doors, in secure facilities, sometimes on military installations to which few have access. So, accessibility is not ATC controllers' strong suit. Consequently, even though practitioner opinions are valued, practitioners are often not accessible to outsiders such as researchers. Inaccessibility and high cost to employ professionals' expertise in research is a significant barrier.

One consequence of the barriers to more direct communications between practitioners and researchers is the slow rate of change. As portrayed in Figure 2.2, the main influence on practitioners in their day-to-day operations are the regulations and operating procedures laid out by the regulators. In the case of military practitioners, these can consist of long-term operational guidelines or specific orders for a given day. Similar regulations and orders constrain the activities of civilian practitioners. It is important to note that FAA regulations and guidance that air traffic controllers must adhere to change very, very, slowly. Changes that were great ideas ten years ago may only now be coming to pass as new methods to perform ATC duties. So, identifying potentially valuable areas for aviation psychology research may be relatively easy compared to identifying an efficient path toward deploying the research results into real-world operations. The necessity to move any changes through the appropriate regulatory agencies and processes inevitably results in substantial delays.

Cultural barriers

One barrier can be the practitioners' bias toward their own domain and resistance to new ideas or new ways of doing things. An operational organization can easily be suspicious of ideas that may challenge the status quo. To some degree this is a natural result of pressures, such as deadlines and internal restrictions driving the internal focus. New ways of doing things would require time and additional effort to implement, and that often is viewed as an unjustifiable expense, even though the long-term savings may justify the new ideas. In addition, any researcher attempting to bring in new ideas will be endorsed and promoted by a trusted member of the operations. But building trust requires time and access, and are often in short supply in both the researcher and practitioner domains. Building a strong bridge between the two communities would require committed and persistent outreach to each other, which is quite a challenge to accomplish.

Another barrier that may be classified as cultural is the very different ways practitioners and researchers approach problems. For a researcher, it is imperative that the problem is put in a broader context to determine the contribution the research makes to the existing body of knowledge. That is why research typically involves painstaking review of existing research literature and a very systematic handling of the research topic. Practitioners have much more direct approach to the problems they are aware of. Although from the practitioners' point of view researchers do not seem to "see the forest for the trees," researchers may hold the same view of practitioners. However, it is apparent that both indeed see a forest, but a different forest: practitioners consider the wider operational context; researchers the wider scientific context. Both are right, of course.

Language barriers

Mutual language barriers are also a significant hurdle for moving research results into operations. Both practitioner and researcher communities have their own specialized jargons and terminologies designed to efficiently communicate with their peers. The language challenge exists in both directions. Just as researchers do not often understand the subtleties of the operational environment, practitioners seldom comprehend the nuances of scientific research. Although specialized terminologies facilitate in-group communication, they hinder cross-group communication. Given the lack of smooth communications between researchers and practitioners, technology often is introduced into a space shaped largely by regulations, immediate operational problems, and/or financial considerations.

Researchers lack understanding of the operational environment and jargon. In any complex operational domain, practitioners develop complex language to communicate with each other. For example, in ATC, this language is referred to as "phraseology" and is a very specific, short-hand way of speaking to reduce the broadcasting time. In this environment, excessive speaking could lead to radio frequency congestion. The phraseology also standardizes key words and phrases so that they are familiar to both pilots and controllers. Phraseology communicates messages in the fewest number of syllables possible to ensure comprehension and therefore to ensure flight safety. Only a well-versed and experienced controller can interpret the thousands of phrases utilized in the complex activities of ATC. A researcher who had not previously performed these duties would be unable to understand the phraseology even while listening to communication recordings offline. Similar system-specific jargons can be expected to occur in any other operational setting. So, researchers need to take this into account when planning or executing a real-world research project.

Practitioners lack understanding of the scientific process and jargon. Practitioners also have difficulties understanding the motivation and jargons commonly used by researchers. Practitioners are not provided automatically with updates on the latest research trends and findings in their field. Many practitioners do not receive, subscribe to, or read human factors journals. There may be many reasons, but certainly one is cost (i.e., journal subscriptions are very expensive). Also, practitioners' time is typically monopolized by the performance of their job. They already have enough raw data to memorize and apply in their daily work. They may not have the luxury to be interested in what they perceive as speculative "maybe-someday-in-the-future" changes or ideas. Consequently, they must rely on the system of regulatory bodies to vet new methods and to disseminate new information. However, as pointed out above, this process is slow to get new technologies and procedures to the field. Although practitioners are intensely trained on the latest *regulatory* changes, of which there are many, they are not trained or informed about the latest research developments.

One consequence of the language barrier is that one's viewpoint is inevitably wrapped around the language one uses. Difference in viewpoints can also introduce challenges when practitioners and researchers try to engage each other in the research process. For example, practitioners' detailed operational knowledge can sometimes induce insufficient bounding of research problems. Not understanding the limitations and constraints of the research process may lead to the studying of too many variables or asking too many questions all at once. This may cause delay or steer the research into problems too complex to attack productively.

Possible solutions to barriers

A root cause of the many barriers presented above stems from the lack of opportunities for practitioners and researchers to interact more freely in order to leverage the expertise in each domain. A number of ways to improve the situation are envisioned below.

Hotline from practitioners to researchers

One of the first possibilities for improving communications would be to provide a direct path for aviation practitioners to identify human factors concerns as they notice them. For example, consider a car mechanic complaining that the car designer has placed an often-replaced component in a location that requires a major disassembly to access. Naturally, the mechanic would think that the engineer has never considered the implications of this placement and has never endured the grief of having to perform this unnecessarily complex task. To deal with such situations, Chevrolet now uses a system called "Techline" where car technicians can call General Motors engineers to offer suggestions to make their cars more service-friendly anytime. After all, who better to offer ways to improve tasks than those experts who do them? Note, however, the workload implications of this solution. Practitioners would have to take the time to articulate and submit their suggestions, and archiving and sorting these suggestions will require manpower too.

Happily, there is one case of a very successful direct line back from aviation operators regarding the safety of flight operations: NASA's Aviation Safety and Reporting System (see http://asrs.arc.nasa.gov). As highlighted on the home page, ASRS is a confidential, voluntary, and non-punitive reporting system open for aviation system operators to report any issue that they think could affect safety. These reports can then be analyzed by researchers for trends and to provide data to fuel research hypotheses. The raw reports and/or processed trends can also be used by regulators and designers to attempt to improve safety. The preliminary work on ASRS began as part of an agreement between NASA and the FAA in 1975 (Reynard, Billings, Cheaney, & Hardy, 1986). Within the first 20 years, ASRS processed more than 338,000 aviation incident reports and issued more than 2,500 alert messages (ASRS Celebrates its 20th Birthday, 1996).

The development and use of similar reporting systems to gather more information from more operational settings should be explored.

Publications for practitioners

Practitioners are not only keen on sharing human factors concerns with researchers, they are also keen on learning about the latest R&D technologies and human factors approaches. So in addition to having a hotline for practitioners to communicate more directly with researchers, researchers could also make more effort to increase practitioner awareness of the cutting-edge developments. But turning scientific literature into operational suggestions, recommendations, and cautions that are understandable to the practitioners would require researchers to be able to bridge the language barrier between the two communities. Involvement of a practitioner in the writing would be a great help in bridging the gap.

Although this idea is very sound, there are many organizational barriers that must be overcome to make it a reality. Writing is presently not in the job description of pilots and controllers, and therefore it must be assumed that they would need to engage in such activity in their own time, without direct compensation. Practitioner-oriented publications (so-called trade magazines) perform a valuable function, but it is not considered meritorious for academics to submit their work to them. Copyright issues also often prevent publication of the same research in multiple outlets, even if the presentation styles and audiences would be quite different.

Mutual immersion

As Figure 2.2 illustrates, communications among practitioners, researchers, and sponsors exist intermittently at best. This is, as discussed above, in part due to the profound differences between the practitioner and researcher backgrounds and home environments. One way to better solidify some of these communication channels might be for practitioners and researchers to spend some time immersing themselves in the other's natural environment.

Researcher immersion in the operational environment

The researcher's role of observing practitioners performing their tasks in the operational environment would be similar to that in a laboratory setting, but there is a very important difference. In laboratory research, the experimental tasks and conditions are designed by the researcher, and the participants' behavior and performance are interpreted against them. In the operational environment, the researcher is likely to have limited understanding of the operational tasks and would find it difficult to identify and interpret critical observable actions. It therefore would be very helpful if researchers could first immerse themselves in the activities they are interested in investigating in order to gain insights into the cognitive, social, and technical aspects of the job demands, organizational expectations, and regulations involved. Allowing the researchers to have hands-on experience with the operational tasks would help them develop a better sense of the problems and issues that practitioners initially face as they first learn the task and later as they try to perform the task at a higher level. The researchers would then be in a much better position to design and conduct research that has more direct operational relevance and payoffs.

Such researcher immersion however would not be possible without the practitioner's help. Practitioners would need to take an active role in mentoring the researchers in the nuances of the operational challenges. Here, the practitioners' participation would far exceed their typical role as just a cooperative research subject limited to trying out new tasks and providing training data for the experimenter.

One example is the experience of one of the authors (a veteran air traffic controller). The researcher, having worked closely with the author, was able to earn the respect of those in the field. This greatly facilitated the other controllers sharing their job knowledge with him. These open exchanges were invaluable even when information was shared sometimes in a "venting" manner. Such exchanges allow a receptive researcher to clearly hear the intricate details confronting the controllers early in the process. The nuances and details conveyed are often times the information needed to help prevent a failed attempt at addressing a complex problem. Armed with a deeper, more insightful understanding, a researcher would be much better able to frame the best possible questions and devise more comprehensive approaches and solutions.

Practitioner immersion in the research environment

Practitioners' immersion in the research process either in the laboratory or in an operational setting would greatly facilitate their acquiring some basic understanding of the research process and thereby their ability to communicate with researchers and to contribute to the research outcome. In fact, some practitioners would welcome professional development opportunities or "career paths" that would allow for entry into the research field after mastery of their craft rather than movement up to the administrative tasks that further remove them from the functions of their trade. Given that practitioners often possess a unique pride in what they do and in the high level at which they perform, the motivation to contribute to improving operations by research is often inherent in such practitioners. For example, one author's motivation to pursue advanced education was due to his desire to improve ATC through knowledge of existing problems and curiosity of the potentials of new technologies. Additional personal experiences from the present authors are provided below to illustrate the great benefits of integrating practitioners into the R&D efforts.

One of the authors (a pilot) was involved in helping to develop a critical part of a training program for a new commercial airline. This author had been involved with a human factors researcher at NASA who had been studying the difficulties associated with using the Flight Management System (FMS) software. The author was able to help identify problems by becoming actively involved in the research and becoming a "student" of cognitive learning models. He also helped recruit line pilot instructors who had experience of training pilots with various FMS backgrounds. Further, he helped the researcher to develop appropriate survey questions, analyze data, and draw relevant conclusions. The knowledge gained by the collaboration provided insights into the training difficulties as well as design issues that instructors had to deal with in order to help the pilots learn the new airplane. Issues and recommendations that were identified included: add foundational knowledge about the FMS systems, provide contextual examples of various FMS usage scenarios, provide opportunities with guided practice to cement learning, avail opportunities for practice and repetition, and apply learning and memory theory to improve training efficiency. The results of the research were not only published but also were presented to the airline manufacturer involved in developing industry-leading training programs. The manufacturer was impressed with the results and asked the author-researcher team to participate in the actual development of the training materials. The author continued to learn and apply lessons from the researcher in courseware development and cognitive learning, while also continuing to be a mentor to the researcher on training issues as well as the task of flying. Ultimately, the research outcome was well received and was incorporated into the training program.

Another example is one of the author's involvement in research developing the VNAV system. As described at the start of this chapter, the use of VNAV itself is complex and is difficult to use effectively even for pilots. For example, VNAV behaves differently depending on the phase of flight, vertical relationship to the programmed vertical path, and external atmospheric variables; ascertaining the state that VNAV is using is challenging, even if the logic is understood by the pilot. A researcher who has not had the opportunity to actually experience its use probably would not be able to identify the technical reasons for the pilots' difficulties. On this occasion, the researcher was developing a tool that would allow a designer to predict gaps in a design and ultimately improve usability and safety. This author was able to mentor the researcher to have a better grasp of the VNAV issues as well as help validate the researcher's hypothesis and modeling concepts. With this assistance, the researcher was able to improve the tool and ensure that the results were in line with expectations from the operating environment. In this case, the depth of knowledge from the author helped bridge the gap between field expertise and research expertise and, acting as a facilitator, he helped to improve the researcher's awareness of confounding variables and dynamics of the operational tasks to develop more accurate and applicable models.

Conclusions

Research priorities for the future should always have the common goal of improved safety. Realistically though, from the industry perspective, any operational changes resulting from research will be driven by increased efficiencies and decreased cost. This is not enough and we need to ask ourselves more visionary questions. What kind of aviation world do we want to create? What role do we want the human to play? What level of risk can we tolerate? What is the role and responsibility of industry, research, and government in this vision? Researchers and practitioners hold the knowledge to help come up with solutions once we know the direction we want to take.

Two areas across the military and civilian sectors emerged from the preceding discussion as needing immediate human factors attention: automation and HMIs. There is perhaps nowhere a more acute need for human factors R&D work than the RPA arena. The insatiable demand for RPAs generated by its suitability for critical military missions will continue to be a growing enterprise within the military. Demand for RPAs is likely to enter the civilian realm as well. Interfaces that enable the RPA operator to effectively control multi-ship/swarm missions with an adequate level of situation awareness and a reasonable level of mental workload must be developed. As noted above, these interfaces also must be able to support sustained operations. A second area of critical concern for aviation psychology is human interaction with increasingly intelligent automation. Automated systems are embedded pervasively in RPA's and NextGen's operations. Automation is counted on to help carry out critical missions in all kinds of terrain and meteorological conditions. The need to optimize HMI is paramount as the synergy of human operators and automation must be robust, and the role of the human as the high level decision maker and director needs to be supported through design innovation.

To better address the real-world needs of aviation practitioners, aviation psychology must consider the combined implications of Figures 2.1 and 2.2. As illustrated in Figure 2.1, aviation psychology, as an applied research domain must combine the best aspects of scientific relevance as well as real-world applicability. This mandates that much improved communication must be established between operators of aviation systems and researchers who strive to improve those systems. Such improved communication will require, but not be fully satisfied by, the two communities simply speaking more to each other. There must be more meaningful involvement of each community into the world of the other. Researchers must strive to understand the subtleties and complexities of the operational environment they wish to affect, and operators must be more involved in the design and execution of research to ensure outcome validity. Both practitioners and researchers recognize the value of being able to spend time together shoulder to shoulder working toward solving a problem or improving a system. It is clear from the panel discussion at the meeting and from the present chapter that many practitioners and researchers are eager to have such opportunities. Institutional and organizational support are not only desirable but very much needed for such opportunities to occur.

Further, Figure 2.2 reminds us that researchers and practitioners are just two of the many stakeholder groups in aviation R&D. For example, practitioners must answer to the regulators that will mandate what equipment and procedures they are to use and follow. For the collaborative effort of the practitioners and researchers to be most effective, they would also need to try to involve the other stakeholders that include research sponsors, designers and engineers, and regulators. In fact, one might add to the picture the organizations that employ practitioners (airline companies and the FAA) and researchers (universities and federal research labs). One possible means of facilitating such interactions might be to make a more concerted effort to include these stakeholders in the practitioner-researcher dialogues at conferences frequented by practitioners and researchers such as ISAP. Of course, much more than occasional meetings at conferences will be needed to make the results achieved there sustainable. But these dialogues may serve as a good start.

References

ASRS Celebrates its 20th birthday (1996). *Callback: From NASA's Aviation Safety and Reporting System,* Number 204, p. 1.

Fitts, P. M. (ed.) (1951). *Human engineering for an effective air-navigation and traffic-control system.* Washington, DC: National Research Council, Committee on Aviation Psychology.

Reynard, W. D., Billings, C. E., Cheaney, E. S., & Hardy, R. (1986). *The development of the NASA aviation safety and reporting system* (NASA Reference Publication 1114). Moffett Field, CA: Ames Research Center.

Sanders, M. S., & McCormick, E. J. (1993). *Human factors in engineering and design* (7th ed.). New York: McGraw-Hill.

Stokes, D. E. (1997). *Pasteur's Quadrant.* Washington, DC: Brookings Institute Press.

Wickens, C. D., Lee, J. D., Liu, Y., & Gordon-Becker, S. (1998). *Introduction to human factors engineering.* Boston, MA: Addison-Wesley-Longman.

Part II

Researcher approaches to support aviation psychology

3 Researcher role in aviation operations

Hans J. Hoermann, Pamela S. Tsang, Michael A. Vidulich, and Amy L. Alexander

This chapter is partly based on the Researcher Plenary Panel held at the International Symposium on Aviation Psychology in May 2015. This Panel followed the Practitioner Plenary Panel held on the preceding day of the symposium. The overarching goal of the two sessions was to foster a dialogue between operational personnel and researchers toward a safer and more efficient flight environment. The charge to the practitioner panelists was to inform the aviation community of their operational challenges. Their thoughts and discussions are captured in Chapter 2 of this volume. The charge to the researcher panelists was to explore best approaches that would bridge the gaps between basic research and current practical applications. The value of use-inspired basic research was discussed to a great extent by Stokes (1997). That use-inspired research would be a good path toward accelerating the process of putting basic knowledge to practical use will be revisited in this chapter.

The chapter begins with briefly presenting the missions of an example research and development (R&D) organization to offer a glimpse of its research operations. Research approaches and practice toward a safer and more efficient flight environment will then be discussed. Two success stories in aviation psychology are outlined as examples of epoch-making research programs in the Pasteur Quadrant. Since collaborations between researchers and practitioners are critical for the success of use-inspired research, means to facilitate their collaborations are explored. Finally, a call for action moving forward will be presented.

Missions of R&D centers

To provide a glimpse of the natural habitat of researchers, the organization and missions of the German Aerospace Center (DLR) will be briefly described. The German Aerospace Center is a large R&D organization with many subunits. It has the mandate to respond to high-level guidance, and aviation psychology research is incorporated into an extensive multi-faceted portfolio that includes many scientific and technological domains.

German Aerospace Center (DLR)

Research institutions in aeronautics or astronautics, such as the DLR or the National Aeronautics and Space Administration (NASA) have to determine the appropriate balance between basic and applied research. This will be decided strategically by stakeholders with regard to the availability of research facilities and workforce capabilities. Due to the culturally and politically diverse conditions in Europe and the European Union, harmonization, interoperability, and mutual alignment have always challenged the definition of any large-scale research agenda such as the Vision 2020 of the European Commission (European Commission, Group of Personalities, 2001).

The DLR has roots going back to the "Aerodynamic Research Establishment" which was founded in 1907 in Goettingen. In the 1960s and 1970s it merged with several other aeronautical research institutions and became the German national research institution for aeronautics and astronautics in 1969, with about 8,000 employees. Its headquarters are located in Cologne. The DLR has the mission to address societal questions on behalf of public customers by conducting research that enhances global mobility and safety while preserving environmental resources. It has committed itself to bridge the gap between basic research and innovative applications, and to transfer knowledge and research results to the industry and political spheres through mediation and consultation as well as through the provision of services. Specific enablers for these strategic objectives are 33 discipline-oriented institutes and contractually regulated national and international partnerships with universities, industry, other research organizations, and the public.

Air traffic has recorded substantial growth worldwide since the mid 2000s, which, in all likelihood, will continue. Such growth cannot be sustained without consequences for the standards and requirements that the air transport infrastructure must meet. For example, ever-increasing urgent questions arise regarding the impact on the environment and climate. Mobility, communication, climate change, demographic development, shortage of resources, safety, and security are among the grand challenges of today. How can research in aeronautics help to enable sustained mobility in a demand-oriented, future-proof and environment-friendly manner?

Key features of DLR's research agenda lie in the holistic consideration of the air transport system as essential for achieving future objectives. International and multidisciplinary collaboration is an indispensable condition to treat these complex questions with greater effectiveness. For example, a long-term cooperation agreement with NASA has recently been renewed. With respect to aviation, this agreement currently includes collaborative efforts in aerodynamics, air traffic management (ATM), and climate research.

The basic structure of DLR's aeronautics program was established in 2007 to align with the European Commission's Vision 2020 (European Commission, Group of Personalities, 2001) and the corresponding strategic research agenda. DLR's main aims are: (a) to increase the efficiency of the air transportation system; (b) to increase the cost-effectiveness of development and operation;

(c) to reduce aircraft noise and harmful emissions; (d) to increase the quality of air transportation for passengers; and (e) to increase safety in the face of growth and external danger.

In this research program, applied industry relevance should also be associated with high scientific aspirations. To this end, the program promotes the following specific application-oriented projects:

- Extended capabilities to analyze and evaluate the overall air transportation system, which includes air traffic, airports, and flight guidance, taking weather and environmental aspects into account. This aims at the performance optimization of the entire ATM system, including its environmental compatibility.
- Development of simulation procedures to support design, evaluation, and certification programs in aeronautics. This aims, for example, at the expanded range of helicopters to all weather conditions.
- Further development of experimental techniques, equipment, and systems for validating technologies and simulation tools in ground and flight tests. The research priorities include, for example, detailed investigations of fossil-based and alternative fuels, reliability, ignition, and combustion stability in commercial jet engines.
- Work for the human-machine interaction in the areas of the cockpit, cabin, and air traffic control work areas, taking psychological and medical aspects into account. This includes human factors research on adaptive pilot and controller assistance, displays, and sensors for improved handling qualities of the aircraft, increased safety, reduced workload, and enhanced situation awareness during the entire operation.
- Research work for a better understanding of the climate impact of air transportation and especially for suitable emission reduction measures in all areas of the air transportation system.

Transcending across all these fields, the research work is supported by a broad range of large-scale facilities—such as wind tunnels, research aircraft, cockpit and tower simulators, combustion chambers and several test facilities for turbines, structures, and materials—as well as the necessary infrastructure for demanding numerical simulations. These facilities provide excellent conditions to pursue research programs in line with international, primarily European, strategies, objectives, and funding measures such as the European Commission's Vision 2020 and Flightpath 2050 (European Commission, High Level Group on Aviation Research, 2011).

Research and practice toward a safer and more efficient sky

Generation versus exploitation of knowledge

For many people, including practitioners and students, the word *science* is often associated with hard study, complicated mathematical equations, voluminous textbooks, and busy scientists wearing white lab coats while talking to themselves. While such impressions may be based on individual experiences, they reflect only a few aspects of science which lean toward the acquisition of new knowledge by basic, laboratory research efforts. In the long run, science is driven by societal and market needs that have been identified and prioritized by policymakers, industries, and research organizations. In this context, science is undeniably applied, solution-oriented, and focused to achieve strategic objectives specified in local, national, or international research agendas. Since societal, market, and environmental conditions are constantly changing, the search for and discovery of new knowledge is an ongoing process to prepare mankind for future challenges. Therefore, as two sides of the same coin, basic and applied science cannot exist without each other. Indeed, successful attempts have been undertaken already in aviation and other domains to establish stronger connections between what is needed by society and what can be provided by science in terms of established knowledge and solutions.

Let us take a closer look at the differences and similarities between the applied and basic approaches. Advantages and disadvantages can be found in both types of approaches along the following dimensions:

(a) Explanation—solution. Basic science reduces the complexity within a real-world domain and generates theories and models that provide explanations for the features and outcomes of a human-machine system under investigation. Applied research strives for procedural knowledge, which can be utilized directly and provides solutions to current operational needs. Because applied research tends to be solution-oriented, the researcher usually is tasked by a customer or a sponsoring body to provide solutions for an existing problem. The sponsor typically has full control of the research question. The time-horizon for a return of the investment is rather short because the application of research findings often can start without delay, dependent only on political decisions. In contrast, results of basic research are usually more abstract and general. Applicability of the findings is not always immediately apparent. General explanations, principles, models, and even identified problems are the expected outcomes from basic research. Because of the generality of the findings, they are also valuable for addressing new issues emerging in the future. Funding for basic research is in most cases provided from public resources such as scientific foundations. The sponsor has less or only indirect influence on how exactly the funding is spent.

(b) Controlled conditions—natural conditions. Basic research is usually conducted within a laboratory environment with tight control of the experimental conditions that affect the observations. Since applied research is oriented toward applicable solutions, much of it takes place within the natural context in which the findings will be utilized.

(c) Scientific rigor—customized. The awareness and control of potential error sources is one of the main quality criteria of basic science. Applied science has to adapt its methods and instruments in accordance with the contextual conditions within the field of application. The dynamics of these conditions follow the natural flow and cannot be fully predicted.

(d) Internal validity—external validity. Because the experimental conditions can be controlled more easily in laboratory environments, results from basic research often are accorded a higher degree of internal validity. This means more conclusive explanations can be formulated and findings are more likely to be replicable under equivalent conditions. However, since certain potentially influential factors in the natural conditions are sometimes intentionally disengaged in the laboratory in order to achieve better experimental control, the transferability of findings from the laboratory to the world outside can be limited. Phenomena demonstrated in the laboratory often cannot be observed in the natural environment.

(e) Frequent publications—infrequent publications. Basic research is primarily conducted in universities and research organizations, where the holy mantra is "publish or perish." In contrast, much of the results of applied research are not published. Although many scientific journals intend to publish more applied research findings in order to attract a wider audience from industry, the pragmatism of applied research is often criticized as not meeting general scientific standards. In addition, applied research results can affect commercial interests of an organization. Therefore, the interest to publish is sometimes overruled by the need of an organization to protect its proprietary knowledge, practice, and technologies.

(f) Reduced risk—elevated risk. The level of risk involved in the execution of basic and applied research differs significantly. Applied research often carries an elevated level of risk because of the higher probability of unexpected side-effects of an intervention or potential interactions with uncontrolled contextual influences. With many factors under control and a lower complexity of possible interactions, the outcome of basic research is more readily foreseeable. But they may carry the risk of not being generalizable beyond the highly controlled laboratory environment.

It is however important to note that these distinctions between basic and applied research are neither dichotomous nor clear cut. In fact, Stokes (1997) argued that research at the intersection of the two approaches where insight and usefulness can be of equal value may be particularly fruitful. Applied and basic research each has its merits and contributes to the development of the other. We provide further evidence of this notion below and advocate leveraging the strengths of each.

Balancing the strengths and weaknesses of applied and basic research

In an attempt to objectively evaluate the relative merits of applied and basic research, Adams (1972) examined the results of two projects. One was the Project Hindsight that was conducted by the United States Department of Defense

(Isenson, 1967; Sherwin & Isenson, 1966). In this project, twenty weapon systems were examined, among them defense systems such as the Minuteman ballistic missiles, the Mark 46 antisubmarine torpedo, and the Starlight Scope for passive night vision. The procedure was to work backward from an important innovation in the system and to ask what R&D events were responsible for it. These R&D events were traced back 20 years to 1945. Over 700 events were found, 91 percent of which could be classified as applied research. In 98 percent of these cases, the investigator was motivated by his awareness that a problem existed, not by pure scientific curiosity or the pursuing of knowledge. Also, 67 percent of the research events occurred before the specific system that it was applied to was begun, with a median time of nine years between occurrence of a research event and its use in a weapon system. In other words, influential research is not short-term research that is done on the system itself, but is often relatively long-term work that occurs well ahead of immediate need. Further, the R&D events were found to have influenced not just a specific weapon system but were applied to a multitude of systems.

The second project was called TRACES (Technology in Retrospect and Critical Events in Science; Loellbach, 1968, 1969). The project examined long-term and basic research events as well as short-term applied ones without any time restriction. Critical R&D events associated with five socially important products: magnetic ferrites, the videotape recorder, the oral contraceptive pill, the electron microscope, and matrix isolation were identified. Over 300 R&D events were found and they were classified into three categories: basic research, applied research, and development and application. As was found in the Hindsight project, most of the applied R&D events occurred 20 years prior to product innovation. But the basic research contributions came even more years before product innovation, peaking 20–30 years before the innovation and mainly preceding applied research. Moreover, basic research was far more influential than applied research. Basic research, applied research, and development and applications were found to be responsible for 70 percent, 20 percent, and 10 percent of the significant events, respectively.

The two projects that Adams described convincingly demonstrated the value and impact of relatively long-term research and thereby the need for researchers to continue to build basic knowledge with rigor. But the time lag between discovery and application could also be significant. Stokes, Adams, and many others recognized the relationship between basic research and applications need not be serendipitous (see also Helton & Kemp, 2011). Allowing practical needs to inform the basic areas that need more intensive research would be expected to reduce the lag.

For example, Gopher and Kimchi (1989) identified three broad human factors topics that could benefit from a better understanding of the underlying psychological principles: visual displays, mental workload, and training of complex skills. With regard to the topic of displays, recent technological developments have afforded practically infinite display format possibilities. Gopher and Kimchi proposed that one principle that should be used to guide display designs is the principle of representation. This is because the most effective display format would probably be the one that is most compatible with how information is represented by the human operator. Therefore, answers to basic questions on how information is represented are likely to benefit many systems and not just a specific display of a single system. With regard to the topic of workload, because increasing computerized automation has been permeating ever more human-machine systems, there is a growing need to be able to know how the system is performing and to be able to predict how the system might perform. Measures of mental workload have been introduced to augment or complement vanishing observable manual responses or complex performance that evades simple quantification (Vidulich, 2003). Gopher and Kimchi proposed that an understanding of the variables that underlie changes in mental workload would be tremendously useful. With regard to training, although automation has certainly increased system capability, it also introduces complexity. This is made plain from the abundance of testimonials from our practitioner colleagues in Chapter 2. Gopher and Kimchi proposed that in order to provide effective and expedient training, we need to have a strong basic understanding of how learning takes place. Instead of reinventing the wheel for each singular training occasion, research on basic principles of skill developments will need to continue.

Importantly, Gopher and Kimchi argued that sound principles should be applicable across a wide range of human-machine systems. For example, a sound training principle should be just as effective training a fighter pilot, an RPA pilot, or an air traffic controller. Further, sound principles should transcend technologies of the day. The existence of unique requirements in each training situation notwithstanding, validated training principles such as the importance of feedback and the incorporation of appropriately challenging elements should apply whether one is learning how to fly a plane with no engines, propellers, or jet engines. Although the years since 1989 have seen great advances in many technologies such as sensors (both environmental and physiological), displays (such as 3D and virtual displays), and especially computer hardware and software (which enable unprecedented levels of automation), the principles that Gopher and Kimchi (1989) referenced are no less relevant today.

Role of our professional organizations. Adams (1972) further suggested that professional organizations could play a more active role. For aviation psychology, this would include such organizations as the Applied Experimental and Engineering Psychology Division of the American Psychological Association, the Human Factors and Ergonomics Society, the Association for Aviation Psychology, and the European Association for Aviation Psychology. All of these organizations could be more proactive about identifying the most profitable topics for basic research that would be likely to support the most effective transition to applied needs. Individual researchers would, of course, still plan their individual research projects, but Adams suggested that a collective vision of research priorities could help to direct efforts to topics that have particular current relevance. And many such topics could be gleaned from the challenges laid out by our practitioner colleagues in Chapter 2. Their insights should serve as fertile ground for identifying knowledge gaps in our human factors and aviation psychology database.

But how can findings from highly controlled basic research be generalizable to the real world? As Projects Hindsight and TRACES have shown, generalizations are not only possible, they are not an anomaly in science. Anderson, Lindsay, and Bushman (1999) considered the belief that laboratory research must have low external validity and field studies must have low internal validity and consequently little hope of bridging the two. They used meta-analytic techniques to examine the consistency of the effects of the same conceptual independent variables on the same conceptual dependent variables between laboratory and field settings across several domains in social psychology (e.g., weapons and aggression, gender and leadership style, age and job-training mastery). Thirty-eight pairs of laboratory and field effects were found based on a literature search of the major psychological journals. Across domains, the correlation of the effect size for laboratory and field studies was .73, a correlation considered to be large by convention (Cohen, 1988). The respectable correspondence between laboratory and field findings showed that the laboratory findings examined must have some external validity and the field findings must have at least some internal validity. Anderson et al. (1999) argued that laboratory research is by no means inherently internally valid and field studies not. Scientifically unsound studies can be conducted in the field as well as in the laboratory. That is, neither internal nor external validity is defined by where the study is conducted, but by the method with which conclusions are drawn.

However, as Chapanis (1988) and many others (e.g., Brunswik, 1955, 1956) have cautioned, generalizations are not guaranteed neither should they be assumed. But, there are a number of standard procedures of scientific methods such as having representative samples, providing sufficient training, and using appropriate measures that would help improve the probability of generalizability. Chapanis also reminded us that one approach to support generalization over a wide range of situations is to purposely design heterogeneity into the studies. That is, the same relationship should be tested over different subjects (e.g., subjects with different levels of experience), tasks, response measures (e.g., decision time and decision quality), and environmental conditions (e.g., whether the human operator interacts with other humans or intelligent agents).

Beyond designing heterogeneity in individual studies, Gopher and Sanders (1984) advocated the back-to-back strategy for the overall research program. While the initial validation of a relationship between certain conceptual variables and dependent measures would mostly be done under tightly control conditions that typically use simple tasks, efforts need to be made to continue to test the relationship with a higher degree of complexity that increasingly approximates that in the target environment to which generalization is to be made. Along the way, results with simple and more complex tasks are compared. This is not only to check the limits of the generality of the relationship being tested. This also affords an opportunity to reveal inaccuracies of the theoretical relationship and possibly ideas for a better representation, thereby contributing to the existing knowledge base.

Last, Wickens and McCarley (Chapter 4, this volume) discussed the potential drawbacks of the conventional overreliance on null-hypothesis significance testing for inferring practical significance. The issue has to do with the overemphasis on trying to avoid the error of incorrectly rejecting a null hypothesis or the error of detecting an effect that is not really present. The risk with this overemphasis is an increased probability of failing to recognize an effect that is in fact present. While this is a concern for both basic and applied research, Wickens and McCarley argue that this may be especially problematic in the applied domains such as those involved with aviation safety. For example, failure to appreciate a true difference between two training methods because the difference did not reach statistical significance at the conventional level of $p < .05$, could lead to the failure to adopt a truly superior training method. While Wickens and McCarley are not advocating abandoning null-hypothesis significance testing altogether, researchers and practitioners are urged to also consider additional approaches, which they describe in Chapter 4 (this volume).

Success stories of research in the Pasteur Quadrant

In this section, two examples of aviation psychological research programs are outlined to illustrate how sustainable the combination of knowledge focused and application focused research could sustainably enhance safety in the air transportation system.

The first example refers to the programs of selection and classification of aircrew members and other personnel for the armed forces (ref. Flanagan, 1947; North & Griffin, 1977; Koonce, 1984; Driskell & Olmstead, 1989; Damos, 2007). The application of psychometric test theory to developing aptitude tests for use in aviator selection has primarily served the purpose of filling jobs with suitably capable people and increasing training effectiveness through reduced washout rates and lower training times. As an applied science, proper psychometric selection leads to substantial economic benefits, improved resources planning, and contributes to better system performance.

Due to the massive demands for suitable pilots and other military personnel during World War II, the development and application of selection tools was incomparably soaring. Huge budgets were allocated in the U.S. to the Air Force programs of Aviation Psychology and Selection and Training of Aircraft Pilots, which were established by the National Research Council in 1939 and 1940. By 1945, these Air Force programs had commissioned a large number of high profile scientists and other professionals who invented new aptitude test batteries, personality inventories, motivation questionnaires, and apparatus tests for aircrew selection. Directed by John Flanagan, these programs produced initially about 30,000 pilots per year (Driskell & Olmstead, 1989), which implies that selection tests had been administered to several hundred thousand applicants by 1945. Based on the unique experience and large data pools from these programs, psychology could advance far beyond selection and training (see Flanagan, 1947, 1948). Groundbreaking new models and theories were developed and became influential in engineering psychology (e.g., Chapanis, Chardner, Morgan & Sanford, 1947; Fitts, 1947, 1951; Roscoe, 1980; Williams, 1980), educational psychology (Gagné, 1965), personality psychology (Cattell, 1950), intelligence models (Guilford, 1956; Cattell, 1971), and human performance taxonomies (Fleishman, Kinkade & Chambers, 1968), to name just a few examples. It is not an overstatement to say that progress in psychology as a basic science and in aviation psychology as an applied science has benefited enormously from this pioneering work in the Air Force Aviation Psychology and Selection programs. Many new R&D tools as well as standards for assessment, training, design, engineering, certification, and management became available and are still highly beneficial for the modern aviation transport system. A still relevant overview of the state-of-the-art can be found in a series of eleven volumes on *Human Factors in Transportation* edited by Barry Kantowitz.

A second, more recent example deals with fatigue management of aviators. Fatigue has been on the NTSB watch list as one of the most serious safety hazards in civil aviation for more than two decades (Rosekind, 2013). Ultra long-range flights, increasing levels of automation and the general public's demand for 24-hour service availability seven days a week have amplified the prevalence of fatigue-related operational errors. As a causal factor for accidents and incidents, fatigue also has been investigated across several other industries, e.g., in ATC—Nealley & Gawron (2015); rail—Bowler & Gibson (2015); maritime—Allen, Wadsworth, & Smith (2008); road transportation—Barr, Popkin, & Howard (2009); firefighting—Dawson, Mayger, Thomas, & Thompson (2015); and medical surgery—Sturm, Dawson, Hewett, & Hill (2011).

Academic psychology carried out many well controlled sleep deprivation studies with hundreds of subjects in laboratories to investigate the extent to which fatigue affects the various aspects of human performance (Lim & Dinges, 2010; Wickens, Hutchins, Laux, & Sebok, 2015). The direct transferability of these laboratory findings to real-world conditions is limited because of the narrowness of the measured skills (e.g., reaction times, memory, or visual attention) and the complexity of underlying factors of sleep regulation. However, these studies revealed important characteristics of physiological processes such as homeostasis, sleep debt, and sleep inertia which became essential elements in bio-mathematical models of fatigue prediction (e.g., Åkerstedt, Folkard, & Portin, 2004).

Applied research was initiated to validate these models in natural settings and to investigate fatigue effects on complex behaviors with higher relevance for the flightdeck environment (e.g., Petrilli, Thomas, Dawson, & Roach, 2006). The problems of small numbers of participants and small numbers of flights, which are often an issue in applied research, could be solved by the collaboration of the Sleep/Wake Research Centre of Massey University in New Zealand and the Sleep and Performance Research Center of Washington State

University in a recent meta-study (Gander, Mulrine, Van den Berg, Smith, Signal, Wu & Belenky, 2015). This epoch-making collaboration combined data from four separate studies with altogether 237 airline pilots on 730 long-range flights between 13 different city pairs and layover times of between 1 and 3 days. With objective and subjective measures of fatigue, this study demonstrated the strong impact of flight timing and flight duration on in-flight levels of sleep propensity and fatigue. It is the first study that could clearly demonstrate the effects of circadian rhythms on sleep propensity, fatigue ratings, and reaction times in a complex operational environment.

Regulatory authorities responded to such findings by adjusting the time length of flight duty periods under different conditions. However, the appropriateness of flight time limitations has been the subject of heated debates ever since they were released, because fatigue can hardly be regulated. As Graeber (2008) emphasized, duty times alone do not manage fatigue risk, because pilots can be overtired even when they are within legal duty times, and vice versa. Cebola and Kilner (2009) hit the nail on the head by asking "When are you too tired to be safe?" on a Eurocontrol ATM R&D Seminar in Munich. Because of the rigidness and arbitrariness of flight duty times, the industry along with joint forces of many stakeholders has proposed to elaborate proactive safety tools that can predict the expected level of aircrew fatigue under foreseeable circumstances. As a result of academic and applied research in combination with practitioners' operational knowledge and industry experience, Fatigue Risk Management Systems (FRMS) became part of ICAO's set of Standards and Recommended Practices in 2011 (ICAO, 2011, 2012). These collaborative efforts enabled a breakthrough in proactive safety work. By systematic application of FRMS, critical elements in flight rosters can be spotted beforehand and flights under elevated risks of fatigue can be avoided. On top of that, when FRMS are being implemented into the operators' Safety Management Systems, the collected data will, for example, drive the refinement of mathematical models of aircrew fatigue and thereby contribute to the generation and validation of new basic knowledge about operators' fatigue in future.

Promoting communications between researchers and practitioners

It is abundantly clear from our practitioner colleagues (Chapter 2) that there is much willingness, even desire, to work with researchers in a number of capacities. The need to work together is equally clear to researchers, and the two success stories described above demonstrate the value of such a beneficial collaboration. But, as discussed in Chapter 2, there are considerable challenges to be overcome to enable better communications. Below are a few avenues for facilitating communications between practitioners and researchers.

Conferences and publications representing cross-sections of practitioners and researchers. Scientific journal publications have long been the staple means for researchers to communicate with each other. Although basic and applied research tended to be published in separate journals historically, that is changing. There is now a growing number of publications with the professed aim of bridging basic and applied research. They include the *Journal of Experimental Psychology: Applied*; *Theoretical Issues in Ergonomics Sciences*; *Human Factors*; *Ergonomics*; *The International Journal of Aviation Psychology*; and *Aviation Psychology and Applied Human Factors*, just to name a few. Periodically, special issues where the entire issue is devoted to a contemporary topic receiving intense attention are put forth in these journals. For example, *The International Journal of Aviation Psychology* has published special issues on pilot selection (1996(2), 2014(1/2)), instructor training (2002(3)), aviation maintenance human factors (2008(1)), synthetic vision (2009(1/2)), and others. These special issues serve as a particularly excellent forum for researchers, airlines, manufacturers, regulators, and service providers to all examine and discuss the "real-world" requirements together with scientifically proven solutions.

Although many of these journals are still primarily written by and for researchers, they all have the requirement that the authors make plain the relevance of the theoretical issues to applications. Some of them also encourage practitioners to submit papers not only to bring operational issues to the attention of researchers but again, to provide a common forum to engage both the researcher and practitioner communities.

Another category of publications are articles primarily written by researchers for practitioners to communicate new findings in a language accessible to them. One example is *Ergonomics in Design*, published by the Human Factors and Ergonomics Society. Certainly, having more publication avenues that have the expressed aims of communicating with practitioners could incentivize researchers to work closely with practitioners in order to produce documents that are written in a language relatable to them.

A fourth category of publications are reports authored primarily by practitioners. The Aviation Safety Reporting System (ASRS, http://asrs.arc.nasa.gov/overview/summary.html (accessed January 13, 2017)) accepts voluntarily submitted aviation safety incident/situation reports from pilots, controllers, and others. The ASRS acts on these reports and identifies system deficiencies, and issues alerting messages to persons in a position to correct them. The authors are not aware of the existence of a similar "hotline" system that would include not just reports of incidents but reports that identify human factors deficiencies or inefficiencies much earlier, prior to the occurrence of incidents. In the present highly networked world, such a reporting system might provide a more direct pathway for connecting practitioners and researchers early on in the problem-solving process.

Even closer communication between practitioners, researchers, and industry representatives is facilitated at conferences and workshops such as the annual meetings of the Human Factors and Ergonomics Society, the thematic lectures of the Royal Aeronautical Society, the biennial International Symposium of Aviation Psychology, and the European Association for Aviation Psychology Conference.

Of note is that in 2014, the European Commission launched the OPTICS project. OPTICS stands for "Observation Platform for Technology and Institutional Consolidation of Research in Safety." It served as a platform for screening ongoing safety-related research and innovation activities in Europe (http://www.optics-project.eu/ (accessed January 13, 2017)). OPTICS organizes workshops and dissemination events once or twice a year. These workshops offer the opportunity to engage with policy and decision makers as well as leading aviation safety researchers to confirm promising research avenues and adjust the ongoing and future safety research agendas. Further activities are initiated to compile a living repository, which traces existing and ongoing research and innovation activities with relevance to aviation safety. This will help to strengthen the accessibility of already generated knowledge and solutions, and to benchmark it against agreed strategic goals and upcoming industry needs.

Technological gatekeeper. Another approach for bringing the research to the practitioners and the practical challenges to the researchers is to develop the role of a technological gatekeeper. Adams (1972) suggested that this gatekeeper would read more of the professional engineering and scientific journals than the average practitioner or administrator and serve as a translator of basic science. Also, the gatekeeper would maintain a wide range of relationships with scientists and technologists outside of his or her organization. While there is likely to be someone in each organization assuming such a role already, Adams (1972) suggested such a role might be formalized and rewarded in order to maximize information transfer.

Organizational climate and institutional support. Institutional support for facilitating the activities described above is indispensable. Many of these activities like the support for publications, publication subscriptions, and conference and workshop participations, will undoubtedly incur costs, but so do ineffectual designs, failed training, and unsafe operations. There are additional essential institutional supports that might entail a paradigm shift in thinking at the operational sites as well as in the academic institutions. For example, there need to be mechanisms and reward structures in the workplace for practitioners to be able to participate in research collaboration much more fully than is currently typical (see Chapter 2). Similarly, academic researchers would need to be able to take certain risks in endeavoring less tightly controlled work that does not necessarily produce data that are acceptable only in theoretical journals. That is, not only would communications between researchers and practitioners need to be improved, government, industry, and academic administrators very much need to be in the loop as well (see Figure 2.2 in Chapter 2).

Research funding scheme. An important lesson from Projects Hindsight and TRACES as well as from our examples is that longer-term research in the end could have a much higher payoff. Consequently, an overemphasis on funding only work that seeks a quick but possibly only a one-time application is unlikely to be the winning strategy. This is especially important in aviation where the time needed to develop and deploy new aircraft or air traffic control systems can require many years. These lessons have made some inroads into many of the major funding agencies in the United States and in Europe. At the same time, major federal funding agencies that have traditionally funded primarily basic research such as the National Science Foundation and the National Institutes of Health, for some time now have required grant proposals to include explicit statements of the broader impact of the proposed research. The European strategic planning effort for the commercial travel system is another good example that seems to have heeded this lesson well.

At the Paris Air Show in June 2001, the Advisory Council for Aeronautics Research in Europe (ACARE) was instituted with over 40 organizations across Europe and representatives of the European Parliament. ACARE was tasked to find consensus on how aviation could better serve society's needs in the future. The result was the "European aeronautics: A vision for 2020" report (European Commission, Group of Personalities, 2001). Since we are presently approaching the year 2020, the vision was revised and published as "Flightpath 2050: Europe's vision for aviation" in March 2011 (European Commission, High Level Group on Aviation Research, 2011). Most of Europe's national and international funding schemes for research in aviation (e.g., Framework Programs, Horizon 2020, SESAR, CleanSky) are based on objectives—basic as well as applied—as defined in these documents.

An important element of ACARE's strategy is to establish a network for strategic research in aviation for the involved stakeholders (including industry, research establishments, academia, regulators, etc.) and to facilitate stakeholder cooperation in Europe and internationally. For example, a number of specialized instruments in the Horizon 2020 and CleanSky programs are currently used to push innovation closer to the market and to stimulate the dialogue between academics, producers, and end users. Thereby, a fresh contract between government and science is made, which will make the case for continued societal investment in realistic terms of the problem-solving capacity of science.

Although much of the European 2020 and 2050 visions are directed well beyond aviation psychology issues such as cleaner aircraft engines and recyclable aircraft, very ambitious goals for aircraft automation were identified which certainly will have a significant impact on the operators' roles and responsibilities. For example, Flightpath 2050 envisioned a future where, "Automation has changed the role of both the pilot and the air traffic controller. Their roles are now as strategic managers and hands-off supervisors, only intervening when necessary" (European Commission, High Level Group on Aviation Research, 2011, p. 9).

The means for achieving those goals, or indeed even proving their viability, will be a tremendous challenge for aviation psychology research and must include careful collaboration between researchers investigating the issues and practitioners that will ultimately need to use the resulting systems.

Forward-looking, future-oriented research

Innovative science does not only depend on available research facilities and financial resources. It is the breed of human curiosity, bright minds, team spirit, enthusiasm, and creativity. These factors are often somewhat fluctuant and can hardly be scheduled by duty-rosters or stringent roadmaps. Ad hoc solutions to current problems are rarely ingenious. Therefore, science policy makers and science managers have to consider that today's investment into organizational climate, individual promotion, and campaigning for young talents will serve our society's interests of tomorrow. Successful science management will have to find the right balance between controlling the expenditure of taxpayers' money and maintaining motivation and creativity-inspiring tolerance for research within the program-oriented funding schemes. This is not to say that we should neglect current market opportunities. It means that basic research even without a focus on immediate utilitarian thinking will serve tomorrow's needs if it explicitly addresses identified or likely trends into future problems. Following the philosophy of Antoine de Saint-Exupéry: "Your task is not to foresee the future, but to enable it" (Saint-Exupéry, 1950, p. 152).

References

Adams, J. A. (1972). Research and the future of engineering psychology. *American Psychologist*, *27*, 615–622.

Åkerstedt, T. Folkard, S., & Portin, C. (2004). Predictions from the three-process model of alertness. *Aviation, Space and Environmental Medicine*, *75*, 75–83.

Allen, P, Wadsworth, E., & Smith, A. (2008). Seafarer's fatigue: A review of the recent literature. *International Maritime Health*, *59*, 1–4.

Anderson, C., A., Lindsay, J. J., & Bushman, B. J. (1999). Research in the psychological laboratory: Truth or triviality? *Current Directions in Psychological Science*, *8*, 3–9.

Barr, L., Popkin, S., & Howard, H. (2009). *An evaluation of emerging driver fatigue detection measures and technologies*. Report FMCSA-RRR-09-005. U.S. Department of Transportation, Federal Motor Carrier Safety Administration.

Bowler, N., & Gibson, H. (2015). *Fatigue and its contribution to railway incidents*. Rail Safety and Standards Board (RSSB) Special Topics Report, February.

Brunswik, E. (1955). Representative design and probabilistic theory in a functional psychology. *Psychological Review*, *62*, 193–217.

Brunswik, E. (1956). *Perception and the representative design of psychological experiments* (2nd ed., rev. & enl.). Berkeley, CA: University of California Press.

Cattell, R. B. (1950). *Personality: A systematic, theoretical, and factual study*. New York: McGraw.

Cattell, R. B. (1971). *Abilities: Their structure, growth, and action*. Boston, MA: Houghton Mifflin.

Cebola, N., & Kilner, A. (2009). When are you too tired to be safe? Exploring the construction of a fatigue index in ATM. *6th Eurocontrol ATM Safety & Human Factors R&D Seminar*, Munich, Germany.

Chapanis, A. (1988). Some generalizations about generalization. *Human Factors*, *30*, 253–267.

Chapanis, A., Chardner, W. R., Morgan, C. T., & Sanford, F. H. (1947). *Lectures on men and machines: An introduction to human engineering*. Baltimore, MD: Systems Research Laboratory.

Cohen, J. (1988). *Statistical power analysis for the behavioral sciences* (2nd ed.). Hillsdale, NJ: Erlbaum.

Damos, D. L. (2007). *Foundations of military pilot selection systems: World War I*. Technical Report 1210. United States Army Research Institute for the Behavioral and Social Sciences.

Dawson, D., Mayger, K., Thomas, M. J. W., & Thompson, K. (2015). Fatigue risk management by volunteer firefighters: Use of informal strategies to augment formal policy. *Accident Analysis and Prevention, 84*, 92–98.

Driskell, J. E., & Olmstead, B. (1989). Psychology and the military: Research applications and trends. *American Psychologist, 44*(1), 43–54.

European Commission, Group of Personalities (2001). *European aeronautics: A vision for 2020*. Retrieved from http://ec.europa.eu/research/transport/publications/items/european_aeronautics__a_vision_for_2020_en.htm (accessed January 13, 2017).

European Commission, High Level Group on Aviation Research (2011). *Flightpath 2050: Europe's vision for aviation*. Retrieved from http://ec.europa.eu/transport/modes/air/doc/flightpath2050.pdf (accessed January 13, 2017).

Fitts, P. M. (1947). *Psychological research on equipment design* (Research Rep. No. 17). Washington, DC: Army Air Forces Aviation Psychology Program.

Fitts, P. M. (ed.) (1951). *Human engineering for an effective air navigation and traffic-control system*. Washington, DC: National Research Council Committee on Aviation Psychology.

Flanagan, J. C. (1947). Research reports of the AAF Aviation Psychology Program. *American Psychologist, 2*, 374–375.

Flanagan, J. C. (ed.) (1948). *The aviation psychology program in the Army Air Forces* (Research Report 1). Washington, DC: U.S. Army Air Forces Aviation Psychology Program.

Fleishman, E. A., Kinkade, R. G., & Chambers, A. N. (1968). *Development of a taxonomy of human performance*. Technical Progress Report 1. AIR-726–11/68-TPR1. American Institute for Research. Washington Office.

Gagné, R. (1965). *The conditions of learning*. New York: Holt, Rinehart and Winston, Inc.

Gander, P. H., Mulrine, H. M., Van den Berg, M. J., Smith, A. A. T., Signal, T. L., Wu, L. J., & Belenky, G. (2015). Effects of sleep/wake history and circadian phase on proposed pilot fatigue safety performance indicators. *Journal of Sleep Research, 24*, 110–119.

Gopher, D., & Kimchi, R. (1989). Engineering psychology. *Annual Review of Psychology, 40*, 431–455.

Gopher, D., & Sanders, A. R., (1984). S-Oh-R: Oh stages! Oh resources! In W. Prinz & A. F. Sanders (eds.), *Cognition and motor behavior* (pp. 231–253). Heidelberg: Springer.

Graeber, R. C. (2008). Fatigue risk management systems within SMS. *Proceedings of the FAA Fatigue Management Symposium: Partnerships for Solutions*, Vienna, VA: June 17–19.

Guilford, J. P. (1956). The structure of intellect. *Psychological Bulletin, 53*(4), 267–293.

Helton, W. S., & Kemp, S. (2011). What basic-applied issue? *Theoretical Issues in Ergonomics Science, 12*, 397–407.

International Civil Aviation Organisation (ICAO) (2011). *Guidance material for development of prescriptive fatigue management regulations*. Annex 6, Part I, Attachment A-1. ICAO, Montreal, Canada.

International Civil Aviation Organisation (ICAO) (2012). *Fatigue risk management systems. Manual for regulators*. Doc 9966 (1st ed.), ICAO. Montreal, Canada.

Isenson, R. S. (1967). *Project hindsight (final report)*. Washington, DC: Department of Defense, Office of the Director of Defense Research and Engineering.

Koonce, J. M. (1984). A brief history of aviation psychology. *Human Factors, 26*(5), 499–508.

Lim, J., & Dinges, D. F. (2010). A meta-analysis of the impact of short-term sleep deprivation on cognitive variables. *Psychological Bulletin, 136*, 375–389.

Loellbach, H. (ed.) (1968). *Technology in retrospect and critical events in science (TRACES).* (National Science Foundation Contract NSF-C535) Vol. 1. Chicago, IL: Illinois Institute of Technology Research Institute.

Loellbach, H. (ed.) (1969). *Technology in retrospect and critical events in science (TRACES).* (National Science Foundation Contract NSF-C535) Vol. 2. Chicago, IL: Illinois Institute of Technology Research Institute.

Nealley, M. A., & Gawron, V. J. (2015). The effects of fatigue on air traffic controllers. *The International Journal of Aviation Psychology, 25*, 14–47.

North, R. A. & Griffin, G. R. (1977). *Aviator selection 1917–1977.* Technical Rep. No. SP-77-2. Pensacola, FL: Naval Aerospace Medical Research Laboratory.

Petrilli, R. M., Thomas, M. J. W., Dawson, D., & Roach, G. D. (2006). The decision-making of commercial airline crews following an international pattern. In: *Proceedings of the Seventh International AAvPA Symposium*, Manly, NSW, Australia.

Roscoe, S. N. (ed.) (1980). *Aviation psychology.* Ames, IA: Iowa State University Press.

Rosekind, M. (2013). Managing fatigue in aviation. *Aviation Safety Coordinators*, NTSB, July 24, 2013.

Saint-Exupéry, A. D. (1950). *The wisdom of the sands.* New York: Harcourt, Brace & Co.

Sherwin, C. W., & Isenson, R. S. (1966, June 30). *First interim report on project hindsight (Summary).* Washington, DC: Office of the Director of Defense Research and Engineering.

Stokes, D. E. (1997). *Pasteur's Quadrant: Basic science and technological innovation.* Washington, DC: Brookings Institution Press.

Sturm, L., Dawson, D., Hewett, P. J., & Hill, A. G. (2011). Effects of fatigue on surgeon performance and surgical outcomes: a systematic review. *ANZ Journal of Surgery, 81*, 502–509.

Vidulich, M. A. (2003). Mental workload and situation awareness: Essential concepts for aviation psychology practice. In P. S. Tsang & M. A. Vidulich (eds.), *Principles and Practice of Aviation Psychology* (pp. 115–146). Mahwah, NJ: Lawrence Erlbaum Associates.

Wickens, C. D., Hutchins, S. D., Laux, L., & Sebok, A. (2015). The impact of sleep disruption on complex cognitive tasks: A meta-analysis. *Human Factors, 57*, 930–946.

Williams, A. C., Jr. (1980). Discrimination and manipulation in flight. In S. N. Roscoe (ed.), *Aviation psychology* (pp. 11–30). Ames, IA: Iowa State University Press.

4 Commonsense statistics in aviation safety research

Christopher D. Wickens and Jason S. McCarley

Imagine a research project to help pilots better understand the flight management system (FMS) modes and respond appropriately to unexpected events. Twenty line pilots from commuter airlines are selected to participate. Ten undergo a conventional training program, and the others go through a novel, augmented program. After a one-week delay, the participants return to the lab for a test of learning, in which they encounter an unexpected configuration of the FMS in a high-fidelity simulator. The time until the initial correct diagnosis and response serves as a measure of performance. Data show a mean response time of 9.5 seconds for the augmented training group and 14 seconds for the control group, a non-significant (*n.s.*) ($p > .05$) effect. A follow up study of 16 participants (8 per group), using a slightly revised version of the augmented curriculum, also produces a non-significant benefit ($M = 3$ seconds, $p > .05$) for the new training program. The researchers conclude, based on the two studies showing no significant benefit, that the new curriculum is no more effective than the standard training program. The developer of the augmented curriculum points out, however, that if the samples of the two studies are pooled, with a resulting N of 18 per group, the mean difference of 4.5 seconds between groups is significant ($p < .05$). Furthermore, the developer notes that the p-values in the original studies were, respectively, .07 and 0.11, both near the "official" .05 cutoff.

This hypothetical but plausible scenario illustrates the potentially serious flaws in classical null-hypothesis significance testing (NHST) as applied in our safety-critical profession of aviation psychology, aerospace human factors, and more broadly in safety-critical human factors research. The likely conclusion by the research sponsor that the augmented training program was ineffective quite possibly resulted in a decision not to adopt it, and perhaps a resulting failure to reduce FMS-related mishaps.

In the following, we will outline some of the concerns underlying the above sequence of events, and suggest remedies. We will draw on some of Wickens' previous thinking about "common sense statistics" (Wickens, 1998), which itself was inspired by an earlier article on aviation safety by Harris (1991), by Loftus (1996), and by Cumming's (2012, 2014) more recent work on "The new statistics."

Four problems with NHST

To begin, we can consider some problems that afflict NHST (though this list is hardly exhaustive, e.g., see Schmidt, 1996; Wagenmakers, 2007).

Problem #1: significance tests give the illusion of a categorical distinction between true effects from non-effects

Conventional NHST practice calculates a probability, *p*-value, for an effect of interest, then compares that *p* to a threshold value alpha (α) to classify the effect as statistically significant ($p \leq \alpha$) or non-significant ($p > \alpha$). The *p*-value itself represents the probability that the data would have shown a result as extreme as the one observed, or more extreme, if the effect of interest were null, that is, if the null hypothesis were true. Thus, an effect is declared statistically significant if the probability of a result as or more extreme than the one observed is less than or equal to α.

The NHST is a largely heuristic process, of course, contaminated by error variance in the data and arbitrariness in the choice of α. Unfortunately, researchers often interpret α as a veridical cutoff between "real" and null effects (Nelson, Rosenthal, & Rosnow, 1986; Dixon, 2003), drawing false certainty from inherently probabilistic evidence. Fisher developed the concept of the *p*-value and recommended $p = .05$ as a criterion for statistical significance, but he regarded *p* as "a measure of the rational grounds for the disbelief" (Fisher, 1959, p. 43) in a hypothesis, not as a touchstone for all-or-nothing judgments (Schneider, 2015). By this way of thinking, we may acknowledge practically or psychologically important benchmarks along the continuum of *p*-values, just as we define 25,000 feet as the "sterile cockpit" threshold on approach, but we need not disregard variations in evidence between or beyond those benchmarks. It was Neyman and Pearson (1933) who rejected the idea of *p*-values as graded evidence, holding that probabilistic findings could never prove or disprove a hypothesis. Instead of trying to decide whether any particular hypothesis is true or not, they argued, we should, "search for rules to govern our behavior . . . [which] insure that, in the long run of experience, we shall not be too often wrong" (p. 291). Neyman (1957) described this as *inductive behavior*, the use of statistical evidence to guide our choice of action, to be contrasted with the idea of inductive reasoning, the use of statistical evidence to guide our beliefs. Hubbard and Bayarri (2003) provide a nice review of the distinction between the two classical statistical approaches.

For scientists who believe it's possible and desirable to reason to conclusions about hypotheses, though, Fisher's philosophy of hypothesis testing is probably more useful than Neyman and Pearson's (Schneider, 2015). In fact, a commitment to inductive behavior is not just unnecessary, but potentially costly. If we analogize statistical analysis to a form of automation, the distinction between Fisher's approach and the Neyman-Pearson approach parallels the distinction between Stage 2 automation (information integration and inference) and Stage 3 automation (decision making). And as pointed out elsewhere

(Parasuraman, Sheridan, & Wickens, 2000; Onnasch, Wickens, Li, & Manzey, 2014), errors of automation at Stage 3 tend to have more dire consequences than those at Stage 2.

In data analysis, the tendency toward dichotomous, significant-or-not, black-white thinking can often lead to "statistical illogic" of the following sort: An analysis of variance (ANOVA) reveals a significant effect of workload across three levels, low (L), medium (M), and high (H), but pairwise post-hoc tests reveal that only the L versus H contrast is significantly ($p < .05$) different. If L differs from H, and M lies in between, how can M be simultaneously "equal" to both L and H?

And beyond just causing such confusion, dichotomous statistical thinking tends to blatantly mislead, in two different ways. Most obviously, it causes researchers to dismiss effects that are real but happen to fall short of statistical significance at the $p = 0.05$ level. More perniciously, it allows a spurious effect that sneaks under the $p = 0.05$ cutoff to live on, in the literature and in application, indefinitely. Having achieved statistical significance once, the "finding" is deemed real, and failures to replicate it are blamed on poor methodology or low statistical power. In truth, a *p*-value in the range of 0.05 is at best tentative evidence of a replicable effect (Cumming, 2008). An effect just shy of $p = 0.05$ should not be dismissed, neither should an effect just under $p = 0.05$ be treated as conclusive.

Problem #2: NHST is biased toward the status quo

Table 4.1 presents the standard decision matrix underlying NHST. Across the top is the "ground truth" state of the world that the researcher wishes to discover. In our example, this is the truth of whether the augmented FMS training program improves flight safety, or does not (the null hypothesis, H_0). We run an experiment to test that possibility, compute statistics, then derive a conclusion based on whether or not our *p*-value falls below α. Our potential conclusions are represented in the rows of the table. Factorially combining the two states of the world by the two potential conclusions, two forms of misjudgment are possible. A *type I* error, in the top right cell of the matrix, occurs when we erroneously conclude there is an effect where in fact there is none. A *type II* error, in the bottom left cell, occurs when we fail to detect an effect that does exist.

In conducting a null-hypothesis significance test, we control our risk of type I error directly, with our choice of α. We control our risk of type II error—or alternatively, our statistical *power*, 1 minus the risk of type II error—indirectly. If we hold α constant, then power is a function of the size of the effect we're studying, and of our sample size, *N*. Since the true effect size is unknown (if it were known, we wouldn't be doing the research), we achieve a desired level of power by estimating the effect size and selecting an appropriate *N*. Although neither Fisher nor Neyman and Pearson endorsed a strict and context-invariant value for α, convention generally allows an α no higher than .05. Power of .80,

Table 4.1 The conventional table of statistical decisions within NHST

		State of the world	
		New training improves safety	*New training does not improve safety*
Experimental results	Disconfirm H_0 ($p < .05$)		Type I error. Strongly discouraged
	Fail to disconfirm H_0 ($p > .05$)	Type II error. Considered more tolerable than a type I error	

however, is considered reasonable (Cohen, 1988), and observed power levels in published research are often lower than that (Cohen, 1962; Sedlmeier & Gigerenzer, 1989). In other words, research practice is willing to accept a false-negative rate four or more times higher than its false-positive rate, overlooking a real effect 20 percent of the time or more. In essence, there is a direct analogy to the criminal justice system, which cares more to avoid punishing an innocent person than freeing a guilty person and hence requires evidence beyond a reasonable doubt for a criminal conviction.

This state of asymmetric concern for type I and type II errors is a natural outgrowth of concern in basic sciences that false-positive discoveries are more costly than false negatives, and obviously, it's counterproductive for researchers in any domain to assert effects that turn out to be untrue. As the recent "crisis of replication" in psychology and other sciences has shown (Pashler & Harris, 2012), non-replicable effects undermine confidence in research, and ultimately make it difficult to convince the government, industry, and the public at large that they should support our studies and trust our claims. But should the balance of concern for type I and type II errors be the same in applied, safety-related research as in basic science? As argued below, it should not be, leading to the third problem.

Problem #3: conventional NHST practice considers values in decision making bluntly and inflexibly

As noted above, neither Fisher nor Neyman and Pearson argued that α should be inflexible. Fisher (1935, p. 13) noted that, "It is usual and convenient for experimenters to take 5 per cent as a standard level of significance," but acknowledged that the choice is subjective and sensitive to context. Neyman and Pearson (1933) argued more specifically that researchers' choice of α should reflect the **costs and benefits** of their potential decisions. Of Type I and Type II errors, they wrote (p. 296), "in some cases it will be more important to avoid the first, in others the second . . . just how the balance should be struck, must be left to the experimenter." Modern NHST practice, unfortunately, generally

Table 4.2 The classic expected value decision matrix

		State of the world	
		New training improves safety [p(H)]	*New training does not improve safety [1-p(H)]*
Consumer's decision	Adopt the procedure	Value of mishaps avoided minus cost of adoption	Cost of adoption
	Do not adopt the procedure	Cost of avoidable mishaps	No cost

ignores this advice, forgetting the cost-benefit analysis, and assumes an α of .05 almost universally. For instance, it holds α at .05 even when a Type II error (e.g., rejecting a true safety enhancement) may be very costly.

To illustrate the problem with this approach, Table 4.2 presents a classic decision table from expected value theory, populated by the specific characteristics of our pilot training example. It is similar in some respects to Table 4.1, but distinct in others. The two possible ground truth states of the world are again shown across the two columns, and the two rows again represent potential decisions. Here, though, these are not the researchers' decisions to reject or accept the null hypothesis, but the research *consumers'* decisions to either implement the new training program or reject it. The consumers' decision is very different from the researchers'. Most importantly, it considers the context-specific costs and benefits of different outcomes, particularly for the two types of decision errors. Rather than simply assuming that type I errors are worse than type II errors, as conventional NHST practice does, it attaches precise payoffs to various decision outcomes, including the costs of adopting the new training program and the costs of avoidable mishaps. With these explicit payoffs in mind, accompanied by some estimate of the effect size under study, the decision makers can select an α-level appropriate to the context, trading off the costs and benefits of type I and type II errors to maximize the expected value of their decision.

Problem #4: NHST does not consider the prior probabilities of the null and alternative hypotheses in decision making

A second feature of the decision matrix in Table 4.2 is the explicit presentation, in the top row, of the *prior probabilities* that either state of the world is true, before we have seen our data. Here, *H* denotes the hypothesis that the new training program improves safety, *p(H)* denotes the probability we ascribe that *H* is true before we have seen our new data, and 1-*p(H)* denotes the probability that we ascribe to the possibility that H is not true. These are termed prior probabilities because they reflect our state of knowledge in anticipation of new data. Having observed the data, we update the priors to produce *posterior probabilities* reflecting our new state of knowledge. Those posteriors in turn become the priors for our next study.

A prior probability is quite different from a *p*-value, which represents the conditional probability of observing a data pattern as extreme or more extreme than we've observed, *given that the null hypothesis is true.* Prior probabilities are fundamental to Bayesian statistics, which treats probability as a measure of belief (Howson & Urbach, 2006). NHST, in the frequentist tradition, does not attach probabilities (other than 0 or 1) to hypotheses and therefore takes no account of earlier evidence for or against an effect when calculating the *p*-value for a given study. In other words, NHST considers statistical results in isolation, not in aggregate. One result is that occasional wildly implausible findings may be treated as creditable (Wagenmakers, Wetzels, Borsboom, & Van der Maas, 2011). A second result, perversely, is that a series of effects all in the same direction and of similar magnitude but all *n.s.* may not be considered as creditable evidence of a phenomenon, as illustrated by the case study at the beginning of the chapter. For researchers working with small numbers of data points, the failure to statistically accumulate knowledge over studies may hinder progress dramatically.

In summary, there are two general points to be made here. First, the consumer of the research, who ultimately decides whether to implement potentially safety-critical procedures, needs more information from the researcher than simply the "reject/don't reject" output of a statistical decision rule, an output which implicitly removes this responsibility from the consumer of the research making policy or design decisions. Second, the application of a binary decision rule without adequate statistical power or consideration of payoffs and prior probabilities produces an inherent bias *against* adopting procedures or equipment that might improve safety.

What is to be done?

Below, we outline two general categories of remedies for this state of affairs; changes to how the researcher should approach experimental design and analysis, and changes to the way data are presented in written reports and articles.

Design and analysis

Increase statistical power. By running more subjects or eliminating sources of unwanted variance, we reduce statistical noise and increase statistical power. Within the NHST framework, we may also boost power by replacing conventional tests with robust statistics (Wilcox, 1998) which protect against the loss of statistical power that results when the parametric assumptions of a conventional test are violated. If we can raise our power to .95 against choice of $\alpha = .05$, we will have eliminated the inherent bias against finding safety-improving effects.

Unfortunately, there are two factors that mitigate against an increase in *N*, in aviation and in many other safety-critical professions. First, it may be extremely difficult to find and recruit highly skilled professional workers to participate in experiments and, with a restricted budget the researcher may well feel lucky to get even a small-to-moderate *N*. University subject pool volunteers are easy to recruit in large numbers, but their knowledge and the generalizability of their performance to aviation professionals or even student pilots is highly questionable. Second, as discussed in earlier writings (Wickens, 2009), the events that may be most safety-critical are those which catch the operator most by surprise, the so-called *black swans* (Taleb, 2007). This would be true, for example, of the unexpected first failure used to estimate the success of the augmented FMS training program in the example above. It is often these unexpected events that compromise safety. Responses to a single, first failure event, of which there can, by definition, be only one per participant, do not afford the luxury of averaging across trials to reduce variance, yet they are unique in their ability to yield worst case response times and detection rates (Wickens, Hooey, Gore, Sebok, & Koenicke, 2009). With statistical power low, and high N experiments difficult to obtain in such safety-critical research, it is important to tolerate a higher published p-level for evidence of an effect, even as we acknowledge that our confidence in the effect is lower.

Formulate an alternative hypothesis. An alternative hypothesis can be explicitly formulated to provide equal footing to the null hypothesis of no effect. Indeed, statistical power calculations require this to be done. However, this is typically done by assuming a standardized effect size at one of a few conventional levels, for example, a "medium" effect size of Cohen's $d = 0.5$. For the applied researcher, it is often more compelling to estimate effects in meaningful performance units. In our example at the beginning of the chapter, the researcher may state that under cases of possible spatial disorientation, in which timely diagnosis of an automation-induced upset is critical, any time savings of greater than three seconds is a worthwhile effect. Hence $H_1 = 3$, while $H_0 = 0$. By augmenting these mean values with an estimate of the standard deviation (e.g., based on earlier studies or pilot data), the researcher can formulate a meaningful, context-specific alternative hypothesis in standardized units as necessary to calculate power. Similar "point estimates" of an alternative hypothesis might be made for the percentage improvement in performance accuracy that results from a particular display or training innovation. In aviation psychology and other branches of human factors, where safety margins can often be defined explicitly in terms of time and distance (separation), such alternative hypothesis point estimations are quite feasible.

Naturally, specifying a point value for H_1 does not eliminate the probability of either a Type I or Type II statistical error. Depending on statistical power, there may be a large range of observed effect sizes (e.g., between 1 and 2 seconds) where researchers cannot confidently assume either hypothesis to be true. In this case, the appropriate conclusion is not acceptance of one hypothesis or the other, but to "withhold judgment" (Loftus, 1996): the gray area between the black and the white.

Use "smart statistics" and planned contrasts (and tolerate them if you are a reviewer). The data in Figure 4.1 provide an imaginary case study of a typical experimental result.

Imagine we compare midair collision avoidance response times for pilots using two different forms of display, under both low or high workload conditions.

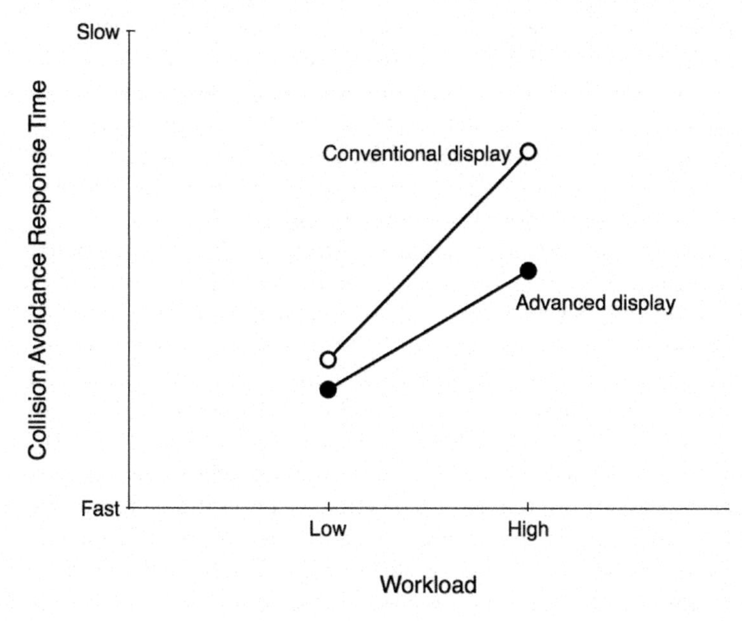

Figure 4.1 Hypothetical results of a 2 × 2 study.

We perform an ANOVA that reveals a statistically significant effect of workload, but an *n.s.* effect of display type and an *n.s.* interaction. Conventional statistics says, "end of story." However, adopting the approach that Wickens (1998) has labeled "smart statistics," we might note that the condition and effect we care most about is collision avoidance under **high workload** (potentially a worst case, safety-critical scenario). Hence what we should do is to focus a **planned comparison** on the high workload condition. For some statistical designs, we may be able to go even further, and test *informative hypotheses* about the specific ordering of means among conditions (Van de Schoot, Hoijtink, & Jan-Willem, 2011). These approaches conform to recommendations from Cumming (2014), who has argued that going into an experiment, investigators should know what they are looking for in terms of specific effects. The omnibus *F*-test is an indirect, inefficient, and sometimes ineffective way of asking whether something specific that we care about is happening. In other words, in the example above, that the advanced warning system can shorten pilot response time in high workload.

A second approach to smart statistical testing in safety research might be to employ one-tailed tests in place of their two-tailed analogues. In using a one-tailed test, safety researchers in effect declare that they care to know if a manipulation improves safety, but don't care to distinguish a safety-compromising manipulation from a null manipulation, since in either case, the proposed safety improvement has failed. This approach may be appropriate for frontline decision makers in industry and government who are responsible for deciding whether to implement new safety-relevant policies or systems. However, one-tailed tests should be used cautiously by academic researchers whose goal is to produce a base of theoretical knowledge to inform safety-relevant practice, as a report of a statistically significant safety cost has very different implications in the scientific literature than a report of a null finding. By employing one-tailed tests that are incapable of acknowledging a safety loss, a scientist may inadvertently hide dangerous manipulations from future researchers.

Replace null-hypothesis tests with parameter estimates, effect size estimates, or model comparisons. A more ambitious solution to the problems inherent in *p*-values is to forego NHST in favor of alternative analytic techniques. The New Statistics movement (Cumming, 2012, 2014) recommends that scientists abandon hypothesis tests and replace them with confidence interval and effect size estimates. Other reformers (e.g., Kruschke, 2015) advocate the use of Bayesian parameter estimates and credible intervals in place of hypothesis tests. Whichever of these approaches we adopt, parameter and effect size estimates shift the focus of analysis from the question, "Are the means different?" to the question, "How different are the means?" A confidence or credible interval can tell the researcher whether or not an effect size is plausibly different from zero, but just as importantly, whether it is plausibly big enough to be practically important. Returning again to the example we started with, a mean improvement in response time of 5.0 seconds with a 95% confidence of ± 1.50 would clearly be of more interest to a decision maker than a mean improvement of 0.05 seconds with a 95% confidence interval of ± 0.015, even if both effects are statistically significant at the same level. The latter effect, even if it's real, is too modest to be of great concern. Effect size and parameter estimates also allow easier meta-analytic accumulation of information (see below) than do dichotomous significant/*n.s.* decisions. Hierarchical estimation methods, which pool information across observers to reduce scatter in the individual estimates (Greenland, 2000), may be especially useful for studies in which large numbers of participants or trials are impractical to obtain.

Hypothesis testing through model comparison offers another alternative to NHST. For example, likelihood ratios comparing model fits can be calculated easily from the output of a standard ANOVA (Glover & Dixon, 2004) or derived from Bayesian techniques (e.g., Rouder, Speckman, Sun, Morey, & Iverson, 2009), and provide more information than null-hypothesis tests in at least two ways. First, by allowing direct comparisons between a null model and alternative models, likelihood ratios allow us to affirm the null hypothesis when it best fits our data. Second, by quantifying the advantage in fit for one model over another, likelihood ratios tell us how much confidence to place in our conclusions. A likelihood ratio that is either very large or very small decisively favors one model over another. A likelihood ratio close to 1.0 is effectively indifferent between models, and tells us that we need more data before we can draw firm conclusions.

Presentation of experimental results

Show the data. It is worth highlighting again the importance of presenting more, rather than less, raw data to research consumers. By raw data, we do not necessarily mean the data points from individual participants (though those may sometimes be appropriate), but rather graphs, confidence intervals, effect size measures, and statistical test outputs other than those of the magical "$p < .05$" type). The added importance of this last bit of guidance to meta-analyses will be described below. Here, again, we can think about statistics and statistical packages in terms of the stages and levels of automation (Parasuraman et al., 2000; Onnasch et al., 2014); late-stage automation is more helpful than early-stage automation when it is correct, but more dangerous when it errs. As noted above, a stats package that simply tells us to accept or reject the null hypothesis is an example of late-stage, decision-aiding automation, and, of course, has some chance of being in error (a 5 percent chance if it is recommending we reject the null, and [1 − power] percent chance if it is recommending that we accept the null). The best mitigation of this, as demonstrated by research on human-automation interaction, is to let the automation provide more assistance in the earlier stage of information integration and inference. In this case, that means providing graphed data and confidence intervals along with the inferential statistics.

Choose language carefully. We should be very careful that the language we use does **not** convey the impression that effects that might be important for safety improvement but fail to reach the magic .05 levels are to be disregarded. Potential offenses here, ranked from bad to worse, might be to describe an effect of, say $p = .07$ with the phrases "not significantly different," "not different," or "equivalent." Even if we report the p-values for such effects, readers who have the time and attention span only for our abstract or discussion may overlook them. More advisable phrasing (although here we have had to argue with editors) would be to label such an effect as "marginally significant," "approaching conventional levels of statistical significance," or as an "*n.s.* trend." Equally important when such effects are in evidence is to describe in the text (not just tables and graphs) their raw magnitudes, in terms such as "a four-second saving in response time," or "a 30 percent gain in accuracy."

Accumulate evidence over experiments. Earlier, we referred to "prior probabilities" for assuming that an effect might actually exist in the world, before we have seen the data from our current experiment. Of course, earlier research is the best source of such priors. Literature reviews can qualitatively summarize that research, but the ideal tool for accumulating evidence over studies is the meta-analysis (Rosenthal, 1991; Borenstein, Hedges, Higgins, & Rothstein, 2009; Cumming, 2014). Meta-analytic approaches provide quantitative estimates of the "collective wisdom" of that prior research, which may enable us to not only know that an effect is likely to be there (or not), but to provide a point estimate of how large it is; that is, an explicit alternative hypothesis. Recognizing the importance of meta-analysis has two implications for us. First, in our own literature reviews, we can use meta-analysis to estimate effect sizes. Second, in reports of our data, we can include the statistical details of our effects including both "significant" and, importantly, "non-significant" effects, with effect sizes given for both, to help other researchers produce unbiased effect size estimates in meta-analyses to come.

Conclusions

P-values and *alpha* have dominated statistical inference in human factors and related fields for decades. But for whatever rigor they've imposed on our scientific reasoning, they have also been used and misused perfunctorily, in ways that have misled us and impeded our progress (Schmidt, 1996; Cumming, 2012), and potentially inhibited advancements in safety. The black-white dichotomy of significant/*n.s.* judgments is likely to have sent many researchers and research consumers chasing spurious effects, and worse, to have smothered or delayed potential improvements to safety. Alternatives to null-hypothesis tests are now within easy reach, and it's possible that many of us will never bother to report a *p*-value again. Even for those of us dedicated to NHST, though, a more careful use of *p*-values and *alpha* is possible, and will improve the quality of human factors research and practice.

References

Borenstein, M., Hedges, L., Higgins, J., & Rothstein, H. (2009). *Introduction to meta-analysis*. Chichester, UK: John Wiley & Sons Ltd.

Cohen, J. (1962). The statistical power of abnormal-social psychological research: A review. *Journal of Abnormal and Social Psychology, 65*, 145–153.

Cohen, J. (1988). *Statistical power analysis for the behavioral sciences*. 2nd ed. Hillsdale, NJ: Erlbaum.

Cumming, G. (2008). Replication and *p* intervals: *p* values predict the future only vaguely, but confidence intervals do much better. *Perspectives on Psychological Science, 3*, 286–300.

Cumming, G. (2012). *The new statistics: Effect sizes, confidence intervals, and meta-analysis*. New York: Routledge.

Cumming, G. (2014). The new statistics: Why and how. *Psychological Science, 25*, 7–29.

Dixon, P. (2003). The p-value fallacy and how to avoid it. *Canadian Journal of Experimental Psychology, 57*, 189–202.

Fisher, R. A. (1935). *The design of experiments*. Edinburgh, UK: Oliver and Boyd.

Fisher, R. A. (1959). *Statistical methods and scientific inference* (2nd ed.). London: Oliver & Boyd.

Glover, S., & Dixon, P. (2004). Likelihood ratios: A simple and flexible statistic for empirical psychologists. *Psychonomic Bulletin & Review, 11*, 791–806.

Greenland, S. (2000). Principles of multilevel modelling. *International Journal of Epidemiology, 29*, 158–167.

Harris, D. (1991). The importance of type 2 error in aviation safety research. In E. Farmer (ed.), *Stress and error in aviation* (pp. 151–157). Brookfield, VT: Avebury.

Howson, C., & Urbach, P. (2006). *Scientific reasoning: The Bayesian approach* (3rd ed.). Chicago, IL: Open Court.

Hubbard, R., & Bayarri, M. J. (2003). Confusion over measures of evidence (p's) versus errors (α's) in classical statistical testing. *The American Statistician, 57*, 171–178.

Kruschke, J. K. (2015). *Doing Bayesian data analysis, second edition: A tutorial with R, JAGS, and Stan*. London: Academic Press/Elsevier.

Loftus, G. R. (1996). Psychology will be a much better science when we change the way we analyze data. *Current Directions in Psychological Science, 5*, 161–171.

Nelson, N., Rosenthal, R., & Rosnow, R. L. (1986). Interpretation of significance levels and effect sizes by psychological researchers. *American Psychologist, 41*, 1299–1301.

Neyman, J. (1957). "Inductive behavior" as a basic concept of philosophy of science. Revue De l'Institut International De Statistique/Review of the International Statistical Institute, *25*, 7–22.

Neyman, J., & Pearson, E. S. (1933). On the problem of the most efficient tests of statistical hypotheses. Philosophical transactions of the Royal Society of London. *Series A: Containing Papers of a Mathematical or Physical Character, 231*, 289–337.

Onnasch, L., Wickens, C., Li, H., & Manzey, D. (2014) Human performance consequences of stages and levels of automation: An integrated meta-analysis. *Human Factors, 56*, 476–488.

Parasuraman, R., Sheridan, T. B., & Wickens, C. D. (2000). A model of types and levels of human interaction with automation. *IEEE Transactions on Systems, Man, and Cybernetics—Part A: Systems and Humans, 30*, 286–297.

Pashler, H., & Harris, C. R. (2012). Is the replicability crisis overblown? Three arguments examined. *Perspectives on Psychological Science*, 7, 531–536.

Rosenthal, R. (1991). *Meta-analytic procedures for social research* (rev. ed.). Beverly Hills, CA: Sage.

Rouder, J. N., Speckman, P. L., Sun, D., Morey, R. D., & Iverson, G. (2009). Bayesian t tests for accepting and rejecting the null hypothesis. *Psychonomic Bulletin and Review, 16*, 225–237.

Schmidt, F. L. (1996). Statistical significance testing and cumulative knowledge in psychology: Implications for training of researchers. *Psychological Methods, 1*, 115–129.

Schneider, J. W. (2015). Null hypothesis significance tests. A mix-up of two different theories: The basis for widespread confusion and numerous misinterpretations. *Scientometrics, 102*, 411–432.

Sedlmeier, P., & Gigerenzer, G. (1989). Do studies of statistical power have an effect on the power of studies. *Psychological Bulletin, 105*, 309–316.

Taleb, N. (2007). *The black swan: The impact of the highly improbable*. New York: Random House.

Van de Schoot, R., Hoijtink, H., & Jan-Willem, R. (2011). Moving beyond traditional null hypothesis testing: Evaluating expectations directly. *Frontiers in Psychology, 2*(24). doi: 10.3389/fpsyg.2011.00024.

Wagenmakers, E.-J. (2007). A practical solution to the pervasive problems of *p* values. *Psychonomic Bulletin & Review, 14*, 779–804.

Wagenmakers, E.-J., Wetzels, R., Borsboom, D., & Van der Maas, H. L. (2011). Why psychologists must change the way they analyze their data: the case of psi: Comment on Bem (2011). *Journal of Personality and Social Psychology, 100*, 426–432.

Wickens, C. D. (1998). Commonsense statistics. *Ergonomics in Design*, *6*(4), 18–22.
Wickens, C. D. (2009). The psychology of aviation surprise: An 8 year update regarding the noticing of black swans. In *Proceedings of the 15th International Symposium on Aviation Psychology*. Dayton, OH: Wright State University.
Wickens, C. D., Hooey, B. L., Gore, B. F., Sebok, A., & Koenicke, C. (2009). Identifying black swans in NextGen: Predicting human performance in off-nominal conditions. *Human Factors*, *51*, 638–651.
Wilcox, R. R. (1998). How many discoveries have been lost by ignoring modern statistical methods. *American Psychologist*, *53*, 300–314.

5 The utility of aptitude in the placement of new air traffic controllers

Cristina L. Byrne and Dana Broach

The U.S. Federal Aviation Administration (FAA) faces a large-scale personnel selection and placement problem in the air traffic control specialist ("ATCS," or more simply "controller") occupation. The generation of controllers hired after the 1981 controller's strike (McCartin, 2011) are reaching mandatory retirement age, and the FAA is hiring the next generation of controllers (FAA, 2015). The FAA projects hiring about 1,000 to 1,200 new controllers per year over the next few years. Tens of thousands of persons apply each year in response to vacancy announcements. With far more applicants than positions, FAA uses pre-employment aptitude testing to winnow the applicant pool down to those with higher potential for success in this demanding occupation.

The next problem for the FAA is assigning this pool of 1,000 or more qualified individuals to more than one job within the organization (Rumsey & Arabian, 2014). Controllers work in one of two general broad types of air traffic control (ATC) facilities. The first type is called "terminal." Terminals handle the beginning and ending phases of flights at airports. Terminal facilities include the iconic Airport Traffic Control Tower (ATCT or tower) and the less well known Terminal Radar Approach Control (TRACON). Controllers in ATCTs provide visually based ATC services to aircraft on the airport surface and in the immediate airspace around the airport up to about 5 miles out. TRACONs provide radar-based ATC services to arrivals and departures up to about 50–60 miles around a single airport or for multiple airports in major metropolitan areas like Chicago. The second type is the Air Route Traffic Control Center (ARTCC, or en route center). Centers provide radar-based ATC services to aircraft en route from one terminal to another. The actual means and methods of work vary by facility type. Therefore, the training process is specific to each type. Before training can begin, the FAA has to assign a new hire to either the terminal or en route training track.

This is a critical decision for the agency in view of the time required for, costs of, and loss rates in ATCS technical training by facility type. Attrition rates vary by training track, with en route training having a historically higher loss rate (about 40 percent) than terminal training (about 20 percent) (Manning, 1998). Terminal training generally requires less time on average (about two years from start to certification) than en route (about three years). Based on data provided by the FAA to Congress (Management of air traffic controller training contracts, 2014), the average cost per year per ATCS trainee to certification was $149,938 in fiscal year 2011. Attrition of an individual from early training might represent a loss to the FAA of less than $100,000, while attrition at two years might represent a loss in excess of $300,000 for a single individual. The economic value of such losses in ATCS training mounts into millions per year, drawing the attention of government "watchdog" organizations and Congress. Moreover, training losses impact facility staffing, leaving positions unfilled, as well as lost time for trainers and trainees. It makes sense, therefore, to try to match an individual to the track in which he or she has a higher likelihood of success from financial, staffing, and operational perspectives.

This chapter focuses on that initial placement decision, that is, the assignment to either terminal or en route training. The chapter is organized as follows. To set the stage, background in the form of a brief orientation to the U.S. ATC system and the controller job is presented along with a description of controller selection and placement for 2006 through 2011, a period of peak hiring. Problems with those processes as highlighted by investigative and stakeholder reports will be described, providing the basis for a testable hypothesis about differential prediction of success by facility type. Empirical testing of that hypothesis is reported in the third section of the chapter. The chapter closes with a discussion of the practical and scientific implications of the empirical results and future research directions.

Background

U.S. ATC system and controller jobs

The core job of an air traffic controller is to ensure the safe, efficient, and orderly flow of air traffic within the assigned area of responsibility. Currently, there are about 14,300 rank-and-file controllers working at 315 FAA air traffic facilities (FAA, 2015). ATC facilities are also classified by level in terms of the volume and complexity of air traffic operations conducted. Volume refers to the number of flights handled on an annual basis by a given facility. Complexity refers to other factors, such as the configuration of an airport, the structure and volume of airspace managed by a center or TRACON, and the types of operations handled by the facility. FAA facilities are assigned to levels 4 through 12, with low volume and lower complexity facilities categorized as Level 4 and high volume and higher complexity facilities categorized as Level 12. For example, Waco ATCT-TRACON is classified at Level 5 (2 runways; 36,138 operations in 2015; little scheduled commercial service); in contrast, Chicago center is a Level 12 facility (2.3 million operations). ATC work at higher level facilities is generally more demanding than at lower level facilities.

ATCS selection and placement, 2006–2011

The FAA and its predecessor agencies have been responsible for the selection, placement, and training of controllers since 1936 (Komons, 1989). There is a large and rich technical literature on U.S. controller selection. However, in this section, we focus on a brief description of the controller selection and placement process used from 2006 through 2011. The FAA began hiring in earnest in 2006 to get ahead of expected retirements from the controller workforce. The pace of hiring dropped off considerably in 2012 and 2013 in the wake of Congressional budget battles and uncertain funding. Moreover, the process has changed considerably since 2014, so we focus on the years of peak hiring. The first step in the selection process in 2006–2011 was an on-line employment application in response to an FAA controller job announcement. Such job announcements generate tens of thousands of applications. The applications are initially screened against specific elements such as U.S. citizenship, age (not older than age 30), and minimum education and work experience. Those applicants that cleared this minimum qualifications hurdle (about 54 percent) were referred for aptitude testing (APT Metrics®, Inc., 2013, p. 23).

AT-SAT computerized aptitude test battery was the second major hurdle in the controller selection process used between 2006 and 2011. AT-SAT was a computerized test battery comprised of eight subtests (Table 5.1). Seven of the eight subtests assessed aspects of cognitive abilities, while one assessed other personal characteristics in the personal history/personality realm. Four of the subtests were dynamic and interactive. The remaining four were static, similar to pencil-and-paper tests. To be clear, AT-SAT was an *aptitude* test battery, where aptitude refers to the abilities and other personal characteristics the person brings to the job and for which the FAA provided no explicit instruction or development. AT-SAT was *not* a test of ATC knowledge or skill.

The eight subtests produced a set of 22 scores ("part scores"); these part scores were weighted and summed with a constant to compute an overall

Table 5.1 Air traffic selection and training (AT-SAT) test battery

Subtest	*Description*
Dials (DI)	Scan and interpret readings from a cluster of analog instruments
Applied math (AM)	Solve basic math problems as applied to distance, rate, and time
Scan (SC)	Scan dynamic digital displays to detect targets that regularly change
Angles (AN)	Determine the angle of intersecting lines
Letter factory (LF)	Participate in an interactive dynamic exercise that requires categorization skills, decision making, prioritization, working memory, and situation awareness
Air traffic scenarios (ATST)	Control traffic in interactive, dynamic low-fidelity simulations of air traffic situations requiring prioritization
Analogies (AY)	Solve verbal and nonverbal analogies that require working memory and the ability to conceptualize relationships
Experience questionnaire (EQ)	Respond to Likert-scale questionnaire about life experiences

composite score on a 0 to 100 scale. The composite was scaled such that a score of 70 on AT-SAT predicted minimally acceptable *job* (not training) performance (Wise, Tsacoumis, Waugh, Putka, & Hom, 2001). The development and validation of AT-SAT is extensively documented (Ramos, Heil, & Manning, 2001a, 2001b).

Persons with AT-SAT composite scores between 70 and 84.99 were placed in the "Qualified" ranking category while persons with scores of 85 or higher were placed in the "Well-Qualified" ranking category (e.g., score bands). Candidates in the "Well-Qualified" category were considered for employment first; when that category was exhausted, candidates in the "Qualified" category were then considered. The placement decision was embedded in the selection decision, as tentative job offers were made for either a specific terminal or en route facility. Selecting officials considered agency need and applicant preference in making the placement decision. There was no formal guidance to selecting officials for using AT-SAT scores in making placement decisions; selecting officials did *not* have access to AT-SAT scores for an individual, only the score band achieved.

As FAA recruitment, selection, and training increased in 2006–2011, the U.S. Department of Transportation Office of the Inspector General (DOT OIG or "Inspector General") began auditing FAA hiring processes with the goal of informing Congress, the FAA, and other stakeholders about FAA's progress in addressing attrition and replacement of retiring controllers (U.S. Department of Transportation Office of the Inspector General, 2009). In a subsequent audit, the Inspector General made an explicit recommendation that the FAA "Evaluate the current AT-SAT test and redesign it so that it results in air traffic controllers being placed at locations according to their skill sets" (U.S. Department of Transportation Office of the Inspector General, 2010, p. 10). A specific concern of the Inspector General was matching individuals to the demands of high volume, high complexity critical ATC facilities such as consolidated TRACONs servicing major metropolitan areas with large hub airports such as New York, Chicago, and southern California. This recommendation presupposed that the "skill sets," (e.g., aptitudes assessed by AT-SAT) vary as a function of facility type and perhaps level.

The placement hypothesis

It seems obvious that aptitude requirements vary as a function of facility type and perhaps level, as suggested by the Inspector General, in view of the differences in the work done by controllers in an ATCT versus an ARTCC. For example, the controllers in a tower can physically see the aircraft, and, in fact, are required to establish a positive identification through correlation of surface radar, visual observation, and pilot report of position on the airport surface. Controllers in a TRACON or an ARTCC do not directly see an aircraft. Rather, they work with a complex, highly symbolic, somewhat abstract two-dimensional representation of aircraft operating in a three-dimensional space.

With the differences in controller roles, pace of operations, complexity, traffic patterns, and control procedures by facility type, it seems natural to conclude that the work performed by controllers varies quite substantially with tower appearing to be very different from radar-based control. Certainly, the ATC-specific knowledge and skills are different by facility type. By extension, it seems reasonable to hypothesize that the aptitudes required by facility type also vary, perhaps substantially. Thus, a logical conclusion would be that these perceived differences in aptitude requirements might be exploited to develop statistical models to predict success in en route versus tower training, resulting in a better match between personal abilities, and facility type and level.

However, job analysis data suggest that, in fact, there are no substantive differences in aptitude requirements by facility type. Specifically, the 1995 job analysis (Nickels, Bobko, Blair, Sands, & Tartak, 1995) found no difference in the importance of the aptitudes required for "doing the work" of a controller by facility type (ATCT versus ARTCC). The job analysts found a high degree of concordance between the profiles of abilities required to do the job in en route and terminal facilities (p. 145). The job analysts concluded that from a selection perspective, there were no substantial differences in "worker requirements" (e.g., aptitudes) by facility type (p. 159).

These two perspectives—the obvious differences in the work by facility type implying differences in aptitude requirements that can be exploited to make placement decisions, and the job analysis finding of no differences in aptitude requirements by facility type—framed the empirical study we report in this chapter. Our goal was to evaluate the Inspector General's recommendation for feasibility, and if feasible, consider the utility of placing controllers into terminal or en route on the basis of aptitude test scores.

Placement on the basis of aptitude

To use any test score for placement purposes, the relevant professional standards and principles require "evidence that scores are linked to different levels or likelihoods of success among jobs" (American Educational Research Association, American Psychological Association, & National Council on Measurement in Education, 1999, p. 160). Relevant evidence might include a pattern of differential relationships between predictors and criteria by job type (Society for Industrial and Organizational Psychology (SIOP), 2003). For example, the correlations between particular AT-SAT subtests and job performance might vary as a function of option (terminal versus en route) or facility level. Other relevant evidence might include differences in expected success and failure by option as a function of aptitude test scores.

Source data

We drew on two validation data sets to evaluate the feasibility of controller placement into terminal or en route training on the basis of aptitude test scores.

We used data from (a) the original AT-SAT validation study for en route (Ramos et al., 2001a, 2001b) and (b) the Concurrent Validation of AT-SAT for Tower Controller Hiring (CoVATCH) project (American Institutes for Research (AIR®), 2012). Each of these studies is briefly summarized.

The en route concurrent, criterion-related validation study was conducted in the late 1990s. Based on the 1995 job analysis, a number of tests were proposed for inclusion in a new computerized test battery. Approximately 1,000 incumbent en route controllers took the proposed test battery for research purposes. Job performance data were collected concurrently in two forms: Behavioral Summary Scale (BSS) ratings of job performance by peers and supervisors; and the en route Computer-Based Performance Measure (CBPM; see Borman et al., 2001). Briefly, the BSS ratings were peer and supervisory ratings of controller job performance across eleven performance dimensions, including, for example, *Maintaining safe and efficient flow of air traffic*, *Adaptability and flexibility* and *Overall effectiveness*. The CBPM, in contrast, was an objective assessment using a situational judgment paradigm. Rather than a verbal description of a situation (say, a conflict between two aircraft), the concept was to emulate the en route controller traffic situation display and present the problems visually and aurally to the controller as a mini-scenario. The traffic situation evolved from starting conditions and then froze. The controller was then presented with a multiple-choice question with response options presenting possible actions to resolve the air traffic problem. The response options had been scaled from least to most effective by panels of expert controllers (Hanson, Borman, Mogilka, Manning, & Hedge, 1999). These two measures (composite BSS score, CBPM score) were combined into a criterion score with a 60 percent weight on CBPM and 40 percent on BSS. The correlation between a weighted combination of scores from the eight tests that became the AT-SAT test battery and the composite job performance measure was .51 without any corrections for range restriction or criterion unreliability. With correction for incidental range restriction, the correlation was .68 (Waugh, 2001).

The concurrent, criterion-related validation study of AT-SAT for the tower was completed in 2012. Incumbent ATCT controllers (N = 302) took the operational version of the AT-SAT test battery. As in the original en route validation study, two measures of job performance were collected: BSS ratings by peers and supervisors; and performance on the Tower Simulation-Based Performance Measure (TSBPM) (see Horgen et al., 2012). The BSS dimensions used in the CoVATCH study were the same as those used in the en route validation, with the behavioral anchors tailored to tower operations. The TSBPM was also based on the situational judgment test paradigm. But rather than a simulated radar screen, the TSBPM simulated the out-the-window view from a tower. As in the en route validation study, the criterion measures were combined into a single composite, with TSBPM weighted at 60 percent and BSS weighted at 40 percent. The correlation between a regression weighted composite of AT-SAT subtest scores and the composite of the two criterion measures was .42 without any statistical corrections (AIR®, 2012). AIR® also found that the regression equation (i.e., the weight given to each subtest) for tower was *not identical* with the equation for en route as reported by Ramos et al. (2001a, 2001b). AIR® concluded that AT-SAT *might* be used for placement by option, but further analyses were needed.

Development of proposed placement approach

AT-SAT scores might be used for placement in many different ways. For example, persons with scores above some cut-off might be assigned to the en route option. Or scores might be categorized into ranges, with persons in the lowest range assigned to one option, persons in the highest range assigned to the other option, and persons with scores in the middle range assigned to either option, depending on agency needs.

The first step in this analysis was to decide how AT-SAT scores might be used for placement. The AIR® researchers suggested computing a score for each option, based on the en route and tower-specific regression equations. The applicants were then assigned to a score band within each option. For example, an applicant could be classified as well-qualified terminal and qualified en route (or vice versa), well-qualified in both, or qualified in both.

The placement procedure suggested by AIR® was feasible but had three drawbacks. First, the overall ranking of an individual (which impacts hiring decisions) was confounded with their ranking within an option (which impacts placement decisions). This might make the initial selection of a candidate more complicated and less systematic with more judgment being required of decision makers for each individual case. For example, a candidate could be Qualified in one option and Well-Qualified in another. If all positions are filled in the option for which they are Well-Qualified, would they be given less consideration for employment, even though they might have been categorized as Well-Qualified overall under previous hiring procedures? A more complicated situation would arise if, for example, an applicant was Well-Qualified or Qualified in one option and Not Qualified in another.

Second, given the width of the categorical bands and the correlation found between the current operational score (representing en route) and the tower score ($r = .65$, see Table 5.2), it would be expected that if the placement rules suggested by AIR® were used, a good portion of candidates would receive the same categorical ranking for both options (i.e., Well-Qualified or Qualified in both options). This would not provide useful information on which to base a placement decision.

Third, the en route equation was reweighted for operational use in 2002 to find an optimal balance between validity and the reduction of adverse impact (i.e., substantially different selection/placement rates which disadvantage protected classes), but the tower equation reported by AIR® was not weighted in a similar way. This suggests that the en route equation used by AIR® in their analysis would produce a different option score for reasons other than "true" subtest relationships to performance. Part of the difference in option scores

Table 5.2 Correlations between AT-SAT scoring variations and first facility success

	Operational	*En route*	*Terminal*
Operational			
En route	.880		
Terminal	.651	.793	
First facility success	.120	.210	.176

Note. All correlations significant at $p < .01$, $n = 2{,}332$; operational, en route, and terminal AT-SAT scores are based on similar, but slightly different equations developed through two AT-SAT validation studies.

would be due to the unique weights composing each option equation, but another part of the difference in option scores could be attributed to the adjustments made to reduce adverse impact that are present in the en route equation and not in the tower equation.

Taking these drawbacks into account, we investigated an alternative approach to placement. The first step would be to categorize individuals using the current operational AT-SAT equation, which was weighted to mitigate adverse impact (Wise et al., 2001), into Well-Qualified, Qualified, and Not Qualified categories using the current cut scores as a basis for initial selection. Second, two additional composite scores would be computed based on (a) the original, unadjusted weights for en route (Ramos et al., 2001a, 2001b), and (b) the tower equation developed by AIR® (2012) for terminal. For convenience, these will be referred to as the *Operational*, *En Route*, and *Terminal* scores, respectively, throughout the rest of this chapter.

The applicant's hiring status would first be determined by using the *Operational* score to determine the initial categorical rankings. Persons categorized as "Not Qualified" on the basis of their *Operational* score would be removed from further consideration. Next, the *En Route* and *Terminal* scores would be computed for each person using the respective option-specific weights. Whichever score was highest would serve as the placement recommendation, as shown in Figure 5.1. In the rare event of a tie, the applicant would be given a recommendation of "*Either*." The initial categorization based on the *Operational* score would then be attached to this option recommendation.

Evaluation of proposed placement approach

The following analyses were conducted to evaluate the proposed placement approach. First, logistic regression analyses were completed to verify the relationship of AT-SAT scores (computed using the three equations) to first facility training success, a binary criterion measure not used in the two previous concurrent, criterion-related validation studies. First facility training success refers to whether developmental controllers achieved certified professional controller (CPC) status at their first facility. Second, cross-tabulations were computed to examine the potential outcomes and utility of using AT-SAT

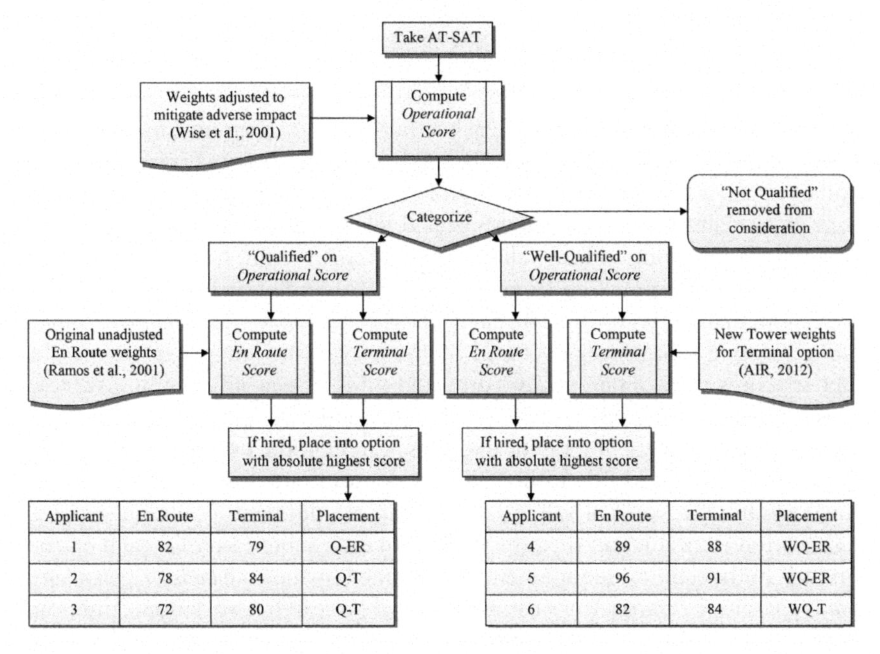

Applicant	En Route	Terminal	Placement
1	82	79	Q-ER
2	78	84	Q-T
3	72	80	Q-T

Applicant	En Route	Terminal	Placement
4	89	88	WQ-ER
5	96	91	WQ-ER
6	82	84	WQ-T

Figure 5.1 Hypothetical ATCS placement process flowchart.

Note. Q-ER = Qualified, *En Route*; Q-T = Qualified, *Terminal*; WQ-ER = Well-Qualified, *En Route*; WQ-T = Well-Qualified, *Terminal*.

for placement. Third, given that placement would constitute an employment decision encompassed by the *Uniform Guidelines on Employee Selection Procedures*, the potential for adverse impact was assessed against the 4/5ths rule (29 C.F.R. § 1607.4D) (Equal Employment Opportunity Commission (EEOC), 1978).

The data used for these analyses were extracted from FAA AVIATOR, the Air Traffic National Training Database, the AT-SAT database, and the FAA Personnel and Payroll System (FPPS). Extracted information included AT-SAT test scores, race, gender, pay, and developmental training status. The sample used for the adverse impact analyses included anyone who had taken AT-SAT ($N = 18{,}663$) *and* who had race/gender information available (Race: $N = 15{,}052$; Gender: $N = 14{,}115$). The sample used for all other analyses included individuals who had AT-SAT data and a finalized first facility outcome (i.e., achieved CPC, failed, or transferred from first facility due to performance) by July 2012 ($N = 2{,}332$). In both samples, individuals had submitted an application for an ATCS position between 2007 and 2009.

Logistic regression analysis

The results of the logistic regression analyses showed that the *Operational*—$R^2 = .022$, χ^2 (1, 2332) = 32.71, $p \leq .001$; *En Route*—$R^2 = .064$,

χ^2 (1, 2332) = 99.32, $p \le .001$; and *Terminal*—$R^2 = .047$, χ^2 (1, 2332) = 71.88, $p \le .001$ scores (based on the previously derived equations) were statistically significant predictors of first facility training success. The raw correlations between these scores and first facility training success, uncorrected for range restriction, were similar but not identical to each other and can be found in Table 5.2. These findings parallel the results obtained during both concurrent validation studies to assess the predictive validity of AT-SAT using ordinary least squares regression analyses and continuously scaled job performance measures, as well as the results of a longitudinal validation of AT-SAT using first facility training success as the criterion (see Broach et al., 2013). Additionally, when logistic regression analyses were run separately for the *En Route* and *Terminal* samples (not restricting subtest weights based on the previously derived equations), the subtest scores were differentially correlated with first facility training success. This is similar to the findings of AIR® that the subtest weights using a sample of tower controllers were not identical to those found in the original en route validation study. Taken together, this evidence demonstrates some degree of differential validity (i.e., overall validity coefficients are different by option samples) and differential prediction (i.e., the weights of each subtest vary for each option sample), both technical requirements as described by the *Standards for Educational and Psychological Testing* (American Educational Research Association, American Psychological Association, & National Council on Measurement in Education, 1999), as well as the *Principles for the Validation and Use of Employee Selection Procedures* (SIOP, 2003). In other words, based on the considerations encompassed by these standards and principles, AT-SAT meets the technical requirements and can be further considered for use in the placement of newly hired controllers into options. These, however, are not the only considerations that need to be weighed by the FAA before using AT-SAT for placement.

Cross-tabulation analysis

The next step in the analysis was to cross-tabulate actual vs. hypothetical placement. "Actual Placement" was the official assignment of newly hired controllers to en route or terminal facilities without regard to their *Operational* score on AT-SAT. "Hypothetical Placement" was the decision that would have been made using the *En Route* and *Terminal* scores derived from AT-SAT. There were four possible combinations of actual-by-hypothetical placements (Table 5.3). The cross-tabulation compared those who were hypothetically placed "correctly" or "incorrectly" in terms of their success in field training (Table 5.4). Placement was "correct" when the actual placement matched the hypothetical placement; placement was "incorrect" when the actual placement did not match the hypothetical placement.

A second cross-tabulation was computed for those who were placed "correctly" or "incorrectly" by their success in field training (Table 5.4), and the results were adjusted based on the typical proportion of hires assigned to

Table 5.3 Actual versus hypothetical placement

Actual placement	*Hypothetical placement*	
	En route	*Terminal*
En route	547 (Correct placement)	297 (Incorrect placement)
Terminal	881 (Incorrect placement)	607 (Correct placement)

Table 5.4 Cross-tabulation of training completion rates at first field facility

Actual placement	*Hypothetical placement*	*Unsuccessful*		*Successful*		*Total*
		N	%	N	%	
En route**	En route	111	20%	436	80%	**547**
En route	Terminal	79	27%	218	73%	**297**
Terminal	En route	153	17%	728	83%	**881**
Terminal**	Terminal	159	26%	448	74%	**607**
Total		**502**	**22%**	**1,830**	**78%**	**2,332**

Note. **Indicates "correct" placement—meaning that applicants were actually placed in the option that AT-SAT would have predicted had it been used for this purpose at the time of hire

each option (Table 5.5). The results of this analysis suggest that the utility of using AT-SAT for placement is marginal to slightly negative. The FAA could potentially see a 3 percent increase to 80 percent in the success rate of controllers "correctly" placed into the en route option as compared to the baseline success rate (without AT-SAT guided placement) of 77 percent. However, this gain could be offset by a 5 percent reduction to 74 percent in the success rate of those "correctly" placed into the terminal option as compared to a baseline success rate (without AT-SAT guided placement) of 79 percent (see Table 5.4) for a net reduction in success rates across both options of 2 percent. However, this loss must be reexamined and weighted within the context of the number of positions available in each option and the number of controllers being hypothetically placed in the en route option. The overall baseline success rate in terminal *without* placement is driven upwards by the higher success rate of individuals that would hypothetically have been placed in the en route option. Given the number of applicants that scored higher on the *En Route* equation, as compared with the number of positions typically available for en route controllers in recent years, it is estimated that approximately 40 percent of available terminal positions *could* be filled by individuals with en route recommendations. This would likely be the preferred policy given their apparent ability to succeed in either option.

Table 5.5 Training success rates at the first field facility with and without placement

	Success rate without placement	*Success rate with placement*
En route (36% of positions)	77%	80%
Terminal (64% of positions)	79%	77%*
Across options (weighted by number of positions)	78.28%	78.08%

Note. *Indicates rate adjusted for likelihood of filling 40 percent of terminal positions with applicants initially recommended for en route placement.

Thus, to accurately estimate the overall success rate *with AT-SAT guided placement* for terminal, given the likely situation that 40 percent of the positions *could* be filled by applicants scoring higher on the *En Route* equation (who would likely have a higher success rate—83 percent vs. 74 percent), a weighted average was computed. The overall success rate for terminal, assuming placement of some applicants with *En Route* placement recommendations into the terminal option, then becomes 77 percent [(83 percent success rate x 40 percent of the positions) + (74 percent success rate x 60 percent of the positions)] instead of the previous estimated success rate of 74 percent for terminal positions. This computation results in a success rate 2 percent lower than the current terminal success rate seen *without using AT-SAT for placement.*

In sum, if AT-SAT is used to guide placement by option, there is a potential increase in success rates for those placed in en route of 3 percent but a potential decrease for those placed in terminal of 2 percent, for an overall 1 percent increase in success rates. However, this estimate must also be considered within the context of the ratio of people hired into each option. Generally speaking, because more people are hired into the terminal option (accounting for approximately 64 percent of open positions yearly), the decrease in the terminal success rate must be weighed more heavily in the calculation of overall success rates computed with and without placement. Taking the higher hiring rate in the terminal option, the net effect of using AT-SAT for placement would likely be a very slight reduction in the overall success rate across both options (Table 5.5).

Adverse impact analysis

As with other employment decisions, a placement decision carries with it the potential to impact an individual's ability to earn. Given the nature of this decision, the potential for adverse impact against members of protected groups must be considered. Using data from FPPS, it was determined that en route controllers earn *on average* approximately $20K more per year than terminal controllers. The difference in annual salaries was calculated using a snapshot of the FPPS data captured in July 2012. This computation produced a very rough

Table 5.6 Adverse impact from placement decision

	Hypothetical placement			*En route placement rate*	*Adverse impact ratio*[a]
	En route	*Terminal*	*Total*		
By ethnicity					
Asian	228	228	456	.50	.95
Black	713	2,324	3,037	.23	**.45**
Hawaiian-Pacific Island[b]	26	49	75	.35	**.66**
Hispanic-Latino	269	556	825	.33	**.62**
Native American-Alaskan Native	30	35	65	.46	.88
White	4,632	4,209	8,841	.52	
Multi-racial	462	569	1,031	.45	.86
No groups marked	357	358	715	.50	.95
Total	**6,717**	**8,328**	**15,045**		
By sex					
Female	1,103	2,320	3,423	.32	**.66**
Male	5,350	5,686	11,036	.48	
Total	**6,453**	**8,006**	**14,459**		

Notes

a Adverse impact ratio calculated with respect to whites for ethnicity and male for sex.

b Groups comprising less than 2 percent of the applicant pool are italicized. Bold ratios are less than what is acceptable under the 4/5ths rule (0.80).

estimate, calculated across all levels of facilities, and is not intended to estimate actual losses or gains an individual controller might experience. Many other factors that could not be measured here would help determine actual losses or gains for each individual. Regardless, *on average*, receiving a recommendation for placement into the en route option would likely provide an individual with a greater *opportunity* to earn more over the course of employment and is, thus, considered the preferred option for calculating adverse impact.

Using the placement rules previously described, assigning controllers to an option using their AT-SAT scores could result in differential placement rates by race and sex into the terminal and en route options (Table 5.6). For example, just 23 percent of black candidates would be recommended for placement in en route, compared to 52 percent of white candidates (adverse impact ratio = .23/.52, or .45, where the threshold for adverse impact is defined as a ratio of .80 or less by the *Uniform Guidelines on Employee Selection Procedures* (EEOC, 1978)). The adverse impact ratio for Hispanic/Latino applicants was .62 and for females was .66.

Conclusions

Looking at both of the AT-SAT concurrent validation studies and this current set of analyses together, there is sufficient evidence to suggest that the abilities required to perform the job of air traffic controller do vary, *to some limited degree*, by option. The regression analyses (calculated repeatedly using different samples and at different times) have, in fact, derived different equations for the two options, which overlap but are not completely identical. This evidence could help provide the technical justification required, if the FAA were to pursue the use of AT-SAT for placement.

However, it is not clear that the variation by option is of a *sufficient degree* to justify differential placement given the minimal utility observed. On the surface, these two options may "appear" more unique than they actually are in terms of the aptitudes required to perform the job. Moreover, the *utility* of using AT-SAT to guide placement is minimal—and might be slightly counter-productive for the FAA. The cross-tabulations indicated that the success rate in en route would increase if AT-SAT were used for placement, but would decrease in terminal. Taken across both options, field training success rates would not likely change in a meaningful way provided that the number of candidates typically hired for each option in recent years remains consistent. Thus, the use of aptitude (at least those measured by AT-SAT) is not likely to provide the FAA with a useful and practical route for the placement of air traffic controllers.

Additionally, in both this study and the AIR® (2012) analysis, the AT-SAT equation derived from a sample of tower controllers was used to represent all of the terminal option, because no data were available examining TRACON-only controllers. Given a similar reliance on radar technology, as well as increased job complexity, ATCSs working in stand-alone TRACONs *might* be more similar to en route controllers than to tower controllers. If the use of AT-SAT for placement purposes were to be further pursued by the FAA, it is recommended that data be collected on TRACON-only controllers to determine whether TRACON-only controllers can, in fact, be placed using the equation derived from tower controllers, or if the en route equation would be more suitable given the similar nature of the work. It might be found that TRACON controllers are more accurately represented using the original en route equation, or that there are substantial differences between all three jobs, and TRACON controllers require an entirely separate equation.

Finally, placement using AT-SAT could potentially have an adverse impact on individuals in protected classes. That is, members of protected classes would be placed into higher paying en route facilities at less than 80 percent of the rate of the majority members of each class (by race and gender). Differential placement rates based on AT-SAT scores could create troubling pay disparities by race and sex. If the FAA were to use AT-SAT for placement, the risk of adverse impact and pay disparities should be evaluated against the marginal utility observed. In sum, given the findings of both validation studies, and the analyses conducted here, using AT-SAT scores to guide placement decisions is not recommended at this time.

Looking forward

As a result of this evaluation of aptitude (as measured by AT-SAT) for use in placement, we are left considering a variety of alternatives. An alternative to looking at general controller aptitude is to focus on measuring more specific aptitudes that might better predict skills required by each option (e.g., radar skill acquisition). One such tool, developed for the assessment of vectoring skill acquisition, has been put forth as a potential placement tool (Baldwin & Hutson, 2015). This assessment is in the initial phase of evaluation for placement purposes.

A secondary alternative under consideration is to broaden the scope of "aptitudes and other personal characteristics" and the outcomes that might be considered in the future. AT-SAT is primarily a cognitive abilities test battery. Outcomes have been restricted to success or failure in training. Personality or work style preferences might predict satisfaction with the initial placement decision and satisfaction with the resulting position upon completion of training. Perhaps the placement question should not be overshadowed by a focus on training completions rates as the outcome of interest, but rather on job satisfaction and commitment to a particular option or facility. Perhaps we need to consider not what these applicants "can do," but rather what they "want to do." Clearly there is much work that needs to be done in regards to the issue of controller placement within the FAA. These decisions are costly for the organization and for the individuals. Often consideration for the individuals is set aside to address organizational concerns, but we advocate for a solution that truly balances these concerns. When newly hired controllers are ill-placed, from their perspective, a new set of organizational problems are created that are also costly. For instance, low satisfaction with placement could lead to increased transfers, decreases in organizational trust and commitment, and understaffed facilities/overworked controllers which can in turn impact a host of issues including performance, as well as labor/management relations (Loi, Hang-yeu, & Foley, 2006; Cropanzano, Bowen, & Gilliland, 2007; Nadiri & Tanova, 2010).

References

American Educational Research Association, American Psychological Association, & National Council on Measurement in Education. (1999). *Standards for educational and psychological testing* (4th ed.). Washington, DC: American Psychological Association.

American Institutes for Research. (2012). Validate AT-SAT as a placement tool. Draft report prepared under FAA contract DTFAWA-09-A-80027 Appendix C. Oklahoma City, OK: Federal Aviation Administration Aerospace Human Factors Research Division (AAM-500).

APT Metrics®, Inc. (2013). *Extension to barrier analysis of air traffic control specialist centralized hiring process. Final report.* Washington, DC: Federal Aviation Administration Office of the Assistant Administrator for Human Resources. Retrieved from https://www.faa.gov/about/office_org/headquarters_offices/acr/eeo_affirm_program/media/Barrier_Analysis_Report.pdf (accessed January 13, 2017).

Baldwin, K., & Hutson, K. (2015). *Radar vectoring aptitude test—Prototype for evaluations.* Report prepared under FAA contract F081-0215BB04-TR. McClean, VA: Mitre Corporation.

Borman, W. C., Hedge, J. W., Hanson, M. A., Bruskiewicz, K. T., Mogilka, H. J., Manning, C., . . . Horgen, K. E. (2001). Development of criterion measures of air traffic controller performance. In Ramos, R. A., Heil, M. C., & Manning, C. A. (eds.). *Documentation of validity for AT-SAT computerized test battery, Volume II.* Report No. DOT/FAA/AM-01/6. Washington, DC: Federal Aviation Administration Office of Aviation Medicine.

Broach, D., Byrne, C. L., Manning, C. A., Pierce, L., McCauley, D., & Bleckley, M. K. (2013, March). *The validity of the air traffic selection and training (AT-SAT) test battery in operational use* (DOT/FAA/AM-13/3). Oklahoma City, OK: Federal Aviation Administration, Civil Aerospace Medical Institute.

Cropanzano, R., Bowen, D. E., & Gilliland, S. W. (2007). The management of organizational justice. *The Academy of Management Perspectives, 21*(4), 34–48.

Equal Employment Opportunity Commission, Civil Service Commission, Department of Labor, & Department of Justice. (1978). Uniform guidelines on employee selection procedures. *Federal Register, 43*(166), 38290–39315.

Federal Aviation Administration. (2015). *A plan for the future: 10-year strategy for the air traffic control workforce 2015–2024.* Retrieved from http://www.faa.gov/air_traffic/publications/controller_staffing/media/CWP_2015.pdf (accessed January 13, 2017).

Hanson, M. A., Borman, W. C., Mogilka, H. J., Manning, C., & Hedge, J. W. (1999). Computerized assessment of skill for a highly technical job. In Drasgow F., & Olson-Buchanan, J. (eds.), *Innovations in computerized assessment* (pp. 197–220). Mahwah, NJ: Lawrence Erlbaum.

Horgen, K., Lentz, E. M., Borman, W. C., Lowe, S. E., Starkey, P. A., & Crutchfield, J. M. (2012). *Applications of simulation technology for a highly skilled job.* Paper presented at the 27th Annual Conference of the Society for Industrial and Organizational Psychology, San Diego, CA.

Komons, N. A. (1989). *Bonfires to beacons: Federal civil aviation policy under the Air Commerce Act, 1926–1938.* Washington, DC: Smithsonian Institution Press.

Loi, R., Hang-yue, N., & Foley, S. (2006). Linking employees' justice perceptions to organizational commitment and intention to leave: The mediating role of perceived organizational support. *Journal of Occupational and Organizational Psychology, 1,* 101–120.

Manning, C. A. (1998). Air traffic controller field training programs, 1981–1992. In D. Broach (ed.), *Recovery of the FAA air traffic control specialist workforce, 1981–1992.* (Report No. DOT/FAA/AM-98/23). Washington, DC: Federal Aviation Administration Office of Aviation Medicine.

Management of air traffic controller training contracts: Hearings before the Senate Subcommittee on Federal Contract Oversight, January 14, 2014 (McNall Questions for the Record (QFR) for Senator McCaskill). Retrieved from http://www.hsgac.senate.gov/download/?id=2B709C2C-DB78-–4B00-–9B7F-325B940B8EF7 (accessed January 13, 2017).

McCartin, J.A. (2011). *Collision course: Ronald Reagan, the air traffic controllers, and the strike that changed America.* New York: Oxford University Press.

Nadiri, H., & Tanova, C. (2010). An investigation of the role of justice in turnover intentions, job satisfaction, and organizational citizenship behavior in hospitality industry. *International Journal of Hospitality Management, 29*, 22–41.

Nickels, B. J., Bobko, P., Blair, M. D., Sands, W. A., & Tartak, E. L. (1995). *Separation and control hiring assessment (SACHA) final job analysis report* (Deliverable Item 007A under FAA contract DFTA01-91-C-00032). Washington, DC: Federal Aviation Administration, Office of Personnel.

Ramos, R. A., Heil, M. C., & Manning, C. A. (eds.). (2001a). *Documentation of validity for the AT-SAT computerized test battery, Volume I.* (Report No. DOT/FAA/AM-01/5). Washington, DC: Federal Aviation Administration Office of Aviation Medicine.

Ramos, R. A., Heil, M. C., & Manning, C. A. (eds.). (2001b). *Documentation of validity for the AT-SAT computerized test battery, Volume II.* (Report No. DOT/FAA/AM-01/6). Washington, DC: Federal Aviation Administration Office of Aviation Medicine.

Rumsey, M. G., & Arabian, J. M. (2014). Military enlistment and classification: Moving forward. *Military Psychology, 26*, 221–251.

Society for Industrial and Organizational Psychology. (2003). *Principles for the validation and use of employee selection procedures* (4th ed.). Bowling Green, OH.

U.S. Department of Transportation Office of the Inspector General. (2009). *Training failures among newly hired air traffic controllers.* (Report No. AV-2009-059). Washington, DC: Author. Retrieved from https://www.oig.dot.gov/library-item/28939 (accessed January 13, 2017).

U.S. Department of Transportation Office of the Inspector General. (2010). *Review of screening, placement, and initial training of newly hired air traffic controllers.* (Report. No. AV-2010–049). Retrieved from http://www.oig.dot.gov/audits?tid=71 (accessed January 13, 2017).

Waugh, G. (2001). Predictor-criterion analyses. In Ramos, R. A., Heil, M. C., & Manning, C. A. (eds.). *Documentation of validity for AT-SAT computerized test battery, Volume II.* (Report No. DOT/FAA/AM-01/6). Washington, DC: Federal Aviation Administration Office of Aviation Medicine.

Wise, L. L., Tsacoumis, S. T., Waugh, G. W., Putka, D. J., & Hom, I. (2001). *Revision of the AT-SAT.* (Report No. DTR-01–58). Alexandria, VA: Human Resources Research Organization.

Part III

Automation and complex systems

6 Visualizing automation in aviation interfaces

Alex Kirlik, Kasey Ackerman, Benjamin Seefeldt, Enric Xargay, Kenyon Riddle, Donald Talleur, Ronald Carbonari, Lui Sha, and Naira Hovakimyan

This chapter considers some of the problems associated with many current approaches to automation design that adopt a displacement approach, in which automated systems are considered as substituting for functions that had typically been performed by humans in system control (e.g., control theory) and human decision making (multi-attribute decision theory and dynamic programming). In contrast, we present two studies illustrating a design approach as an alternative to displacement, one in which automation is used to drive, not vehicles themselves, but instead dynamic visual representations that provide additional information to the human operator on the otherwise opaque dynamics of the vehicular systems being controlled (e.g., visualizing aircraft control safety envelopes or trade-offs associated with competing objectives when optimizing the routing of aircraft on taxiways). These dynamic visual representations are intended to highten operator engagement and to empower more effective and robust decision making and control.

First, we discuss the historical and theoretical context of our approach to these design problems in human factors and aviation psychology.

Undesirable consequences of a displacement approach to automation design

The presence of automated control systems in aircraft is ubiquitous. As demands for aircraft safety and efficiency have increased, so too have levels of complexity found in these systems. While this automation has resulted in significant safety benefits, increased incidents and accidents due to a lack of pilot engagement, variously described as the "out-of-the-loop" (OOTL) problem (Endsley & Kiris, 1995) or "out-of-the-loop unfamiliarity" (OOTLUF) problems (Wickens & Hollands, 2000), have prompted much recent research, including a recent study on automation-induced, task-unrelated thoughts or "mind wandering" by pilots (Casner & Schooler, 2014). This issue has recently achieved renewed levels of recognition from those working in, and reporting on, advanced concepts for the design of increasingly autonomous vehicles (cars), such as the prototype designs recently created and road tested by Google and Tesla. A general consensus has only recently been emerging that design concepts that enable fully autonomous driving yet expect that the driver will be able to rapidly and effectively jump back into the control loop to "save the day" after significant periods of attention placed elsewhere, are unlikely to succeed or at least achieve broad consumer acceptance in the near term (Lafrance, 2015).

Ironically, increasingly reliable automated systems can introduce new threats to safety. This is because the behavior logic of highly reliable automation becomes hidden (opaque, due to normally successful operation over long time periods), thereby leading to pilots that become "surprised" (i.e., have low situation awareness) when they become thrown back into the loop to attempt recovery from rare but inevitable automation failures.

When pilots are required to reenter the control loop unexpectedly, their ability to do so effectively is often compromised, a phenomenon known as "automation surprise" (Billings & Woods, 1994) or the "return-to-manual-control deficit" problem (Hadley, Prinzel, Freeman, & Mikulka, 1999). Various attempts have been made to cope with related problems such as "mode confusion" and "mode error" (Sarter & Woods, 1995; Degani & Heymann, 2002) which result from pilots having an inadquate understanding of automation due at least in part to the fact that automation behavior is insufficiently revealed or presented in cockpit interfaces.

We believe that in large part, these difficulties can be attributed to the fact that there is a significant loss of situational awareness surrounding automation state. While a pilot may be fully aware of various flight variables (such as heading, altitude, or airspeed) that appear on the primary flight display, they have few indications of automation state apart from an active/inactive marker. This setup disregards the fact that highly complex automation contains huge amounts of system information regarding not only current aircraft state, but also the control corrections necessary to maintain that state and potential future states. These automated systems can be seen as "silent co-pilots", controlling the plane but offering no insights into their process. D. A. Norman's (1990) paper on automation sets up a thought experiment where the reader is asked to compare flying with the aid of an automated system with a human flight crew. At the point of failure, Norman describes how the "informal chatter" in the human-only cockpit facilitates early detection of flight problems, while the automated system silently compensates until a more dramatic failure occurs. Overcoming problems of decreased awareness of automation then becomes a problem of reintroducing this informal chatter into the automated cockpit. Pilots should be given a steady stream of non-intrusive information to allow a continuous monitoring of automation's contribution to achieving safe flight.

In addition to control augmentation, automated decision aids are playing an ever-increasing role in the Next Generation Air Transportation System, NextGen (Wickens, Mavor, Parasuraman, & McGee, 1998; Erzberger, 2004). Optimization algorithms are increasingly being proposed as decision aids to help bring system efficiency to a level not achievable by humans alone, or to perform tasks that humans simply do not have the cognitive resources to perform effectively. This is a seemingly prudent approach because, as

Rasmussen (1986) describes, humans are more effective at skill and rule-based behavior than knowledge-based behavior, which is to be avoided in dynamic control tasks whenever possible.

However, the use of such algorithms almost always necessitates automating the more advanced functions of *decision selection* or *action implementation* (Parasuraman, Sheridan, & Wickens, 2000), or otherwise reducing the level of human involvement in the decision process. The inappropriate implementation of advanced automation can result in declines in human performance (Norman, 1990; Sheridan, 2002; Parasuraman & Wickens, 2008), loss of situation awareness (Endsley, 1993; Wickens, 2008), and other issues with human-automation interaction (Parasuraman, 1997). Additionally, optimization algorithms can produce results ranging from sub-optimal to catastrophic if the assumptions on which they are based are not met or some factors that determine safe and effective system operation are not included in the optimization algorithm. In complex dynamic environments such as air traffic control, rarely is it the case in which all constraints are known and all assumptions are met. A human that is simply executing algorithm-derived aiding instructions or acting in a passive monitoring role is ill equipped to handle off-nominal situations for which the algorithm is not effective.

Using automation to enhance visualization

We are not the first to provide approaches to the design of automation interfaces. Degani and Heymann (2002) presented a technique for formal verification of these interfaces using computer science techniques. Their focus was to ensure that interface information is sufficient to specify the meaningful states and state transitions an operator needs to be aware of in order to successfully use automation for system control. Jamieson and Vicente (2005) present an approach to the design of effective human-automation-plant interfaces drawing upon a variety of concepts and resources from control theory. Bennett and Flach (2011) extend and elaborate on an interface design approach grounded in control theory, based on concepts such as controllability and observability, and provide a number of illustrative applications.

In other cases, approaches to the design of automation interfaces have been motivated by an even more general technique called ecological interface design, or EID (Vicente & Rasmussen, 1992). Although the theoretical foundations of EID may be consistent with control theory (see Flach, 2017), this display design approach is most frequently presented using a richer array of constructs, such as Jens Rasmussen's "abstraction hierarchy," "decision ladder," and other formalisms that have no direct, one-to-one mapping to control theoretic concepts per se, or at least on a quantitative basis (see Vicente & Rasmussen, 1992). For example, as described by Burns (2013), "The priorities of EID are to provide complete information, show constraints in a work environment, show the information as visually as possible, and provide support for different work behaviors" (p. 566).

Central to both the control theoretic and ecological approaches to interface design is creating a *functional* model (or models) of the controlled or automated systems for which displays must be created, and revealing this functional information at the interface through the effective design of perceptually friendly graphical forms. For example, Seppelt and Lee (2007) used an approach motivated by EID to design and evaluate a display for visualizing adaptive cruise control automation in automobiles that maps vehicle stopping and following distances into a trapezoidal or triangular display form. The emergent feature of this display is a transition between a triangle and trapezoid shape to reflect a threshold change in distance between the two vehicles with respect to automation functionality. This geometrical display is augmented with dashed boundaries indicating the available limits of the automatic cruise control given road conditions and stopping power. Rather than simply providing raw data to the driver, this design focuses on exploiting and presenting the information emergent in the functional relationships between the driving situation and control automation behavior, and its capabilities and limitations.

Perhaps the most sustained and sophisticated efforts toward developing an ecologically oriented approach to interface design in aviation have been those of the Delft Ecological Design group (Borst, Flach, & Ellerbroek, 2014; Borst, Mulder, & Van Paassen, 2015). As in the research reported in this chapter, this group and their collaborators have addressed both cockpit and traffic control interface design from an approach grounded in functional models of the systems being controlled, along with the design of novel, graphical forms for communicating this functional information. Examples that are most relevant to our own research are presented in Amelink, Mulder, Van Paassen and Flach (2005) and in Mulder (2014). We encourage readers to compare the ecological interface design research of the Delft Ecological Design group with our own, functionally oriented, visualization-based approach to automation interface design as illustrated by the studies described in the following.

Study 1: Visualizing departure scheduling decision automation

An alternative to optimization-based decision aiding automation, which may not be robust to violations of assumptions, is a visualization of the constraints on effective human decision making and action selection (Kirlik, 1995). This approach is conceptually related to ecological interface design, as described by Vicente and Rasmussen (1992) and Burns and Hajdukiewicz (2004). Kirlik, Walker, Fisk, and Nagel (1996) state that, due to time pressure and complexity, humans often opt for heuristic solutions to dynamic decision-making tasks. Knowing that humans will naturally adopt perceptual heuristics, presenting key system constraints in a manner that can be easily processed allows the decision aid design to positively influence the development of these heuristics. Presenting information such that relevant system variables can be included in perceptual heuristics can leverage and amplify the natural processes of skilled dynamic decision making, rather than requiring the operator to turn over part or all of the decision-making process to automation that may not be equipped to deal with all situations.

The purpose of this study was to directly compare the two general aiding approaches of optimization-based automation used for decision aiding versus a visualization of key constraints on action selection. Importantly, the constraints that are visualized are identical to the constraints represented in the multi-objective optimization algorithm used as the basis for the decision aiding automation. Represented as an objective function, the automated decision aid approach is one that attempts to minimize differences between scheduled and actual departure times and is used to find an optimal departure schedule. This schedule is presented to the human controller as guidance. In contrast, the visualization approach visually presents the constraints used in the optimization automation. The human controller then uses his or her knowledge of the goals of the task to attempt to schedule departures in the timeliest manner possible with respect to the constraints on acceptable departure schedules as presented on the system interface. In this manner, rather than providing a "point solution" to the controller that may not be robust to information unavailable to automation, the human operator can see what constraints exist that should be satisfied, but is free to bring any other information to bear on determining an effective departure schedule in addition to those visually depicted. It was hypothesized that this style of decision support would be more robust than the point solution method, especially when information known to the operator yet unknown to automation could result in either an infeasible or inefficient departure schedule.

Study design

This study compared three different versions of a prototype interface designed to improve the efficiency of departure sequencing at Dallas-Fort Worth International Airport (DFW): (1) Baseline (no decision aiding); (2) Temporal Constraint Visualization (TCV); and (3) Optimal Timeline Display (OTD), or simply the "Timeline" display. These decision aids were motivated in part by prior research utilizing algorithms to aid ground controllers in sequencing the release of departure aircraft (Hoang, Jung, Holbrook, & Malik, 2011; Jung et al., 2011).

Simulated DFW surface operations

This simulation operated on a simplified model of south-flow surface operations on the east side of DFW. The surface of the east side of DFW features three terminals, with *hold spots* in front of each terminal. The concept of operations in this simulation was for participants to hold departure aircraft at these hold spots (after pushing back from the gate) until the aircraft could taxi directly to the runway. This was in contrast to the standard practice of sending departures immediately to the runway and creating a backup at the departure queue.

By avoiding backups at the runway departure queue, fuel consumption could be reduced by minimizing stop and go traffic. In addition, controllers could better sequence departures according to scheduled departure time, because aircraft would be sent to the runway in the order they were scheduled to depart, rather than the order they pushed back from the gate. The release of aircraft from hold spots was to be controlled with the goals of avoiding any unnecessary stops during taxi, departing aircraft on time, and maintaining maximal runway throughput when necessary.

Method

Participants

Twenty-one instrument-rated pilots controlled simulated aircraft in a re-creation of surface operations at DFW. Though none had prior experience as air traffic controllers controlling actual aircraft, limiting participation to only instrument-rated pilots ensured that they had an adequate understanding of aviation terminology and a basic understanding of airport surface operations. Training on the nature of the task to be performed, and on the use of the aids when present, was provided to all participants during initial experimental sessions.

Equipment

The experiment was conducted in the Beckman Institute Illinois Simulator Laboratory's flight simulator, which was modified to support an air traffic control experiment. Three projectors displayed a 150 degree simulated out-the-window view from the DFW East Tower and a 70 inch LCD monitor displayed the prototype interface (see Figure 6.1). Participants interacted with the interface via a standard computer mouse and a six-button keypad.

Software

The simulation software automatically controlled the movement of aircraft outside the participants' area of control based on predefined scripts and controlled the movements of aircraft inside the participants' area of control based on participant-issued instructions. Simulation scenarios were generated based on data collected at DFW for the Surface Operations Data Analysis and Adaptation (SODAA) Tool (Brinton, Lindsey, & Graham, 2010). These SODAA data aggregate actual flight information such as call sign, aircraft type, scheduled departure/arrival times, departure/arrival airports, assigned gate, and so on, and link them to positional data collected for each flight. The information contained in SODAA data allows for a dynamic re-creation of aircraft movements based on actual operational data.

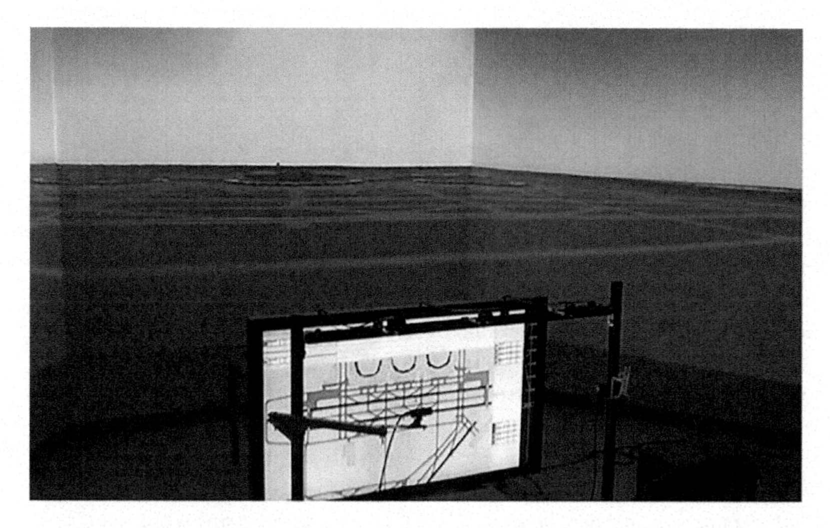

Figure 6.1 The Illinois Simulator Laboratory flight simulator, as configured for this experiment.

Experimental tasks performed

Departure aircraft task. Participants' primary task was to sequence the release of departure aircraft from hold spots to depart aircraft on time and minimize unnecessary Taxi Delay. Each departure aircraft had a scheduled departure time, which appeared on each aircraft's flight strip, and the participants were instructed to use this information when making sequencing decisions. The participants' goal was to have each departure aircraft depart as close to its scheduled time as possible, while minimizing the total taxi time and number of stops in the departure queue.

Arrival aircraft task. Upon reaching the participants' area of control, arrival aircraft would automatically be handed-off to the participant and wait for further taxi instructions. Each arrival aircraft had an assigned terminal and gate, and it was the participants' responsibility to direct the aircraft to the appropriate terminal. Participants needed to keep up with incoming arrivals to prevent a backup of arrivals from potentially blocking a runway. Crucially, and intentionally, the optimization automation did not take into account constraints or other information about the need for the human controller to interleave arrival aircraft on the various surface taxiways with the departure aircraft that were the key element and focus of the ground controller's task. As such, this experimental manipulation made it possible that, in at least some cases, it would be necessary for any good controller to depart from the point solution departure schedule determined by the multi-objective optimization automation aid. This experimental manipulation was motivated by the fact that, in reality, no automation designer can fully anticipate every constraint that could be relevant to decision quality, necessitating human controllers to combine information or task demands known only to them with information also known by automation in decision making.

Interface design

The prototype graphical interface (see Figure 6.2) included the following features—electronic flight strips, flight strip organization windows, an overhead map display, and preset taxi route buttons.

Electronic flight strips. The electronic flight strips contained all of the information needed by participants to effectively control an aircraft. There were nine data fields in each flight strip—including call sign, aircraft type, Estimated Time of Departure (ETD)/Estimated Time of Arrival (ETA), arrival airport, assigned terminal and gate, updated departure time, assigned runway, assigned altitude, and the departure or arrival navigation fix. In addition, the left side of each flight strip included a window with a colored background and black symbol called the *indication box*, used to indicate the current taxi status of that aircraft.

Flight strip organization windows. The electronic flight strips were contained within four flight strip windows appearing on the interface. Two departure flight strip windows contained flight strips for aircraft waiting at hold spots and taxiing to the runway. Two arrival flight strip windows contained flight strips for aircraft arriving on the two arrival runways. Participants could select a specific aircraft for interaction by clicking on its corresponding flight strip. The arrangement of flight strips in the organization windows, along with the indication box for each flight strip, allowed participants to determine the taxi status and approximate position of each aircraft.

Overhead map display. The overhead map display depicted the layout of the airport surface in a similar fashion to current Airport Surface Detection Equipment, Model X (ASDE-X) displays (see Smith, Evers, & Cassell, 1996). A triangle icon showing its current position, heading, and movement represented each aircraft. These icons were colored to match the color of the box on that aircraft's corresponding flight strip (arrival, departure). In addition to selecting aircraft by clicking on the flight strip, participants could also select an aircraft by clicking on its map icon.

Preset taxi route buttons. Participants issued taxi instructions to aircraft by using preset buttons displayed on the interface. These contained the possible routes from the aircraft's current location and could be activated by using the mouse or keypad. By providing preset routes, the participants were spared the mental effort of generating their own turn by turn routes on an unfamiliar airport surface.

Decision aids

The same general interface shown in Figure 6.2 was used for each experimental condition, with the addition of an automated decision aid in the two decision aid conditions.

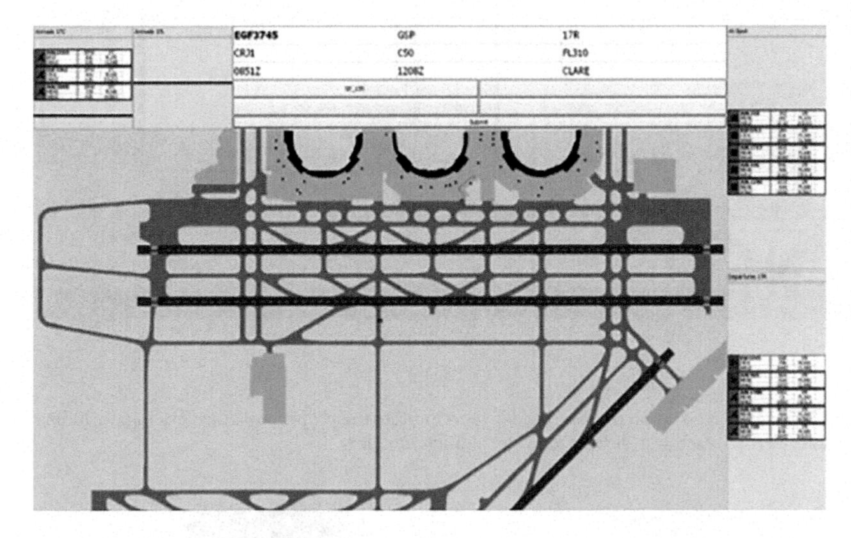

Figure 6.2 The Baseline prototype interface.

TCV aid functionality. The TCV decision aid provided participants with a real-time graphical representation of departure timing constraints, such as the fact that there was a minimum spacing of 45 seconds between departures, and in-trail distance constraints for aircraft of various sizes, as colored lines. These lines allowed participants to perceive minimum aircraft separation in terms of taxi following distance as green lines. In addition, *overlaps* (instances of aircraft spaced too closely) and *gaps* (instances of aircraft spaced farther than necessary) were depicted graphically to provide feedback to participants. This information was used to indicate the earliest location in the departure sequence that an aircraft could be released from its hold spot without incurring any unnecessary Taxi Delay (see Figure 6.3).

OTD or "Timeline" aid functionality. The Timeline decision aid utilized an algorithm to present a suggested departure sequence to the participant. This algorithm functioned by considering the aircraft currently waiting at hold spots with the four earliest scheduled departure times. All permutations of release order and all possible release times for each order were checked via an exhaustive search for the sequence that best minimized total Departure Time Deviation and unnecessary Taxi Delay. The output of this algorithm was a specific release time for each of those four aircraft, which was then presented graphically to the participants (see Figure 6.4), updated every four seconds.

Experimental design

A between-subjects manipulation of decision aid type was performed, with each of the twenty-one participants randomly assigned to one of the three

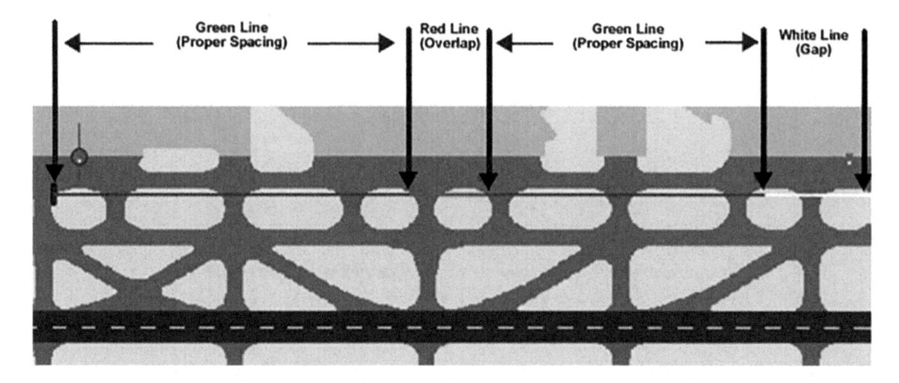

Figure 6.3 The TCV decision aid.

Note. A line was overlaid on the overhead map to indicate the proper departure spacing, including gaps and overlaps; each was differentiated by a different color.

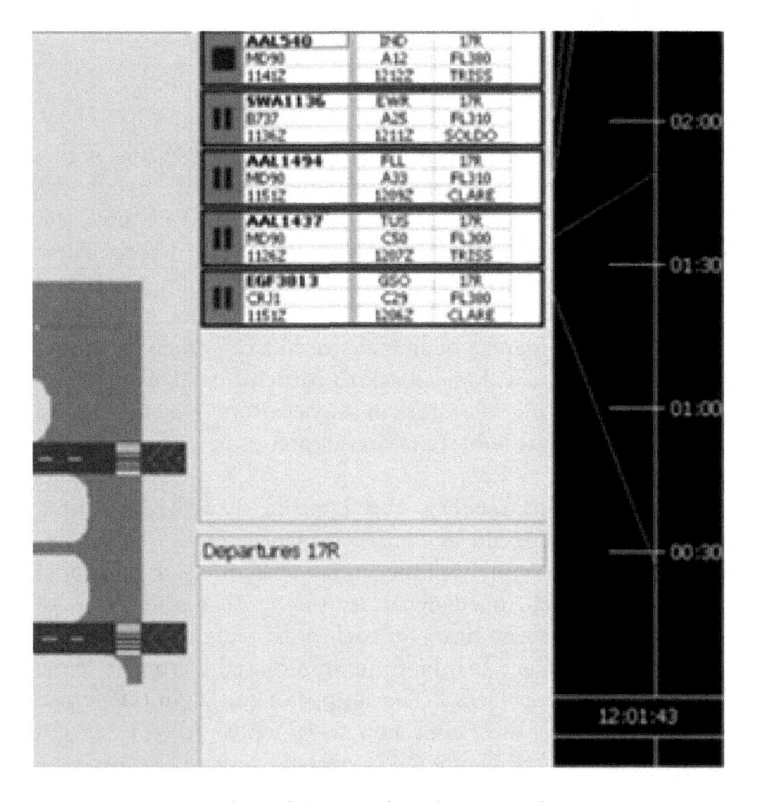

Figure 6.4 A screenshot of the Timeline decision aid.

Note. Lines connect flight strips for aircraft waiting at hold spots to points on a moving timeline to indicate the time at which each aircraft should be released from its hold spot: a point solution departure schedule.

display conditions. After training, eight data-collection trials were performed by each participant, for a total of fifty-six trials per condition. The scenarios for the data-collection trials, as well as the training trials, were presented to each participant in the same order.

Procedure. All participants took part in three experimental sessions lasting between 90 and 120 minutes each. The first session consisted of training and three practice trials, with the second and third sessions consisting of four data-collection trials each.

Training. Participant training consisted of a Microsoft PowerPoint presentation presented on the 70 inch LCD monitor, which was narrated by the experimenter using a script. During the presentation, a paused out-the-window scene of DFW was shown on the projectors and was used by the experimenter as a reference to link certain aspects of the presentation to the out-the-window view. Participants were first given an overview of DFW surface operations and what role they would be playing in the simulation. They were then instructed on the benefits of carefully sequencing the release of departures and how this could be implemented at DFW. Once a high-level overview of the concept had been explained, participants were then provided with in-depth details of how they would perform the tasks required of them using the prototype interface. There was no training variation between conditions, other than simply explaining the functionality of any decision aid that may have been present.

Practice trials. After the completion of the training presentation, participants performed one ten-minute and two twenty-minute practice trials. During the first trial, the experimenter provided active assistance and answered any questions. After that, participants received no more feedback and the two twenty-minute trials were performed in the same manner as data-collection trials.

Data-collection trials. During both the second and third experimental sessions, participants completed four twenty-minute data-collection trials. At the end of each trial, participants were allowed to take up to a five-minute break.

Results: efficiency of departure sequencing

Two measures were used to assess the efficiency of participants' departure sequencing, with both approximating unnecessary fuel consumption. These measures included Taxi Delay and Queue Stops. Taxi Delay was defined as any additional taxi time for an aircraft (in seconds) caused by waiting in the departure queue. Queue Stops were the number of times an aircraft came to a stop inside the departure queue. Without having to wait for an aircraft in front of it, any departing aircraft would taxi directly onto the runway without stopping or slowing down in the departure queue. Therefore, any observed Taxi Delay or Queue Stops were a direct result of inefficient departure spacing and "perfect" performance would be represented by values of zero on both measures.

Taxi delay. A repeated measures ANOVA was performed to test for differences in mean Taxi Delay between the Baseline (M = 36.10, SE = 0.78),

TCV ($M = 5.19$, $SE = 0.23$), and Timeline ($M = 15.27$, $SE = 0.32$) conditions (see Figure 6.5). A significant main effect was found for decision aid type, $F(2, 18) = 4.87$, $p < .05$. A pairwise comparison between display conditions was performed using Tukey's HSD procedure to control family-wise Type I error. There was a significant difference between the Baseline and TCV conditions, $p < .001$, indicating that participants in the TCV condition were more effective at minimizing aircraft taxi time. No significant differences were found between the Baseline and Timeline conditions, $p = .13$, or between the TCV and Timeline conditions, $p = .59$.

Queue stops. A repeated measures ANOVA was performed to test for differences in the mean number of Queue Stops for each aircraft between the Baseline ($M = 0.87$, $SE = 0.012$), TCV ($M = 0.27$, $SE = 0.004$), and Timeline ($M = 0.63$, $SE = 0.006$) conditions (see Figure 6.6). A significant main effect was found for decision aid type, $F(2, 18) = 6.52$, $p < .01$. A pairwise comparison between display conditions was performed using Tukey's HSD procedure to control family-wise Type I error. There was a significant difference between the Baseline and TCV conditions, $p < .01$, and a marginally significant difference between the TCV and Timeline conditions, $p = .10$. These results indicate that aircraft controlled by participants in the TCV condition had fewer stops in the departure queue than both the Baseline and Timeline conditions. No significant difference was found between the Baseline and Timeline conditions, $p = .34$.

Results: timeliness of departures

Participants were assessed on their ability to depart aircraft as close to the scheduled departure times as possible by recording Departure Time Deviation. This was defined as the absolute value difference (in seconds) between an aircraft's scheduled departure time and its actual takeoff time.

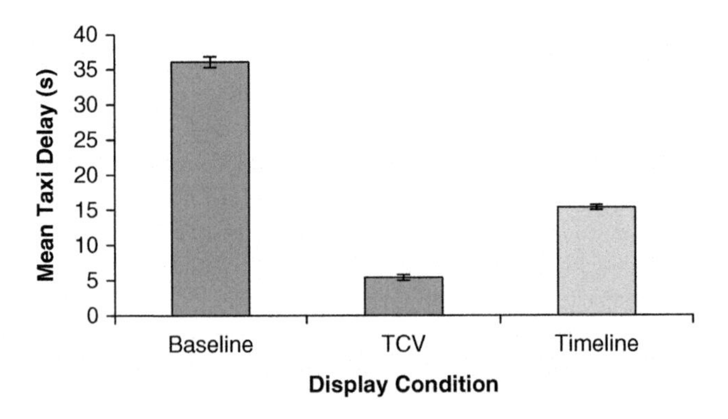

Figure 6.5 Mean Taxi Delay (per aircraft) for each condition, with error bars representing *SE*.

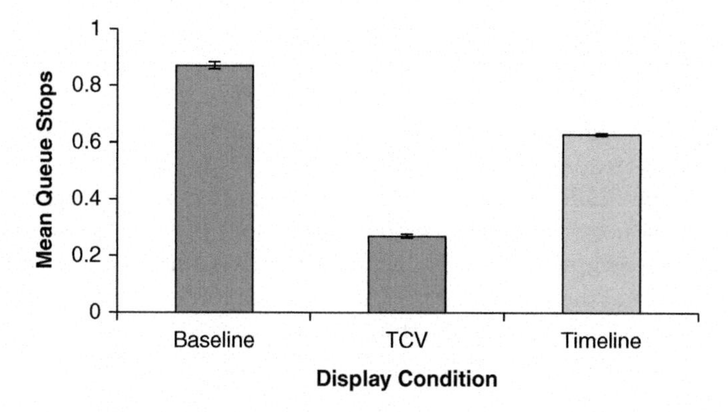

Figure 6.6 Mean queue stops (per aircraft) for each condition, with error bars representing *SE*.

A repeated measures ANOVA was performed to test for differences in mean Departure Time Deviation between the Baseline (M = 188.66, SE = 1.17), TCV (M = 183.66, SE = 1.23), and Timeline (M = 178.10, SE = 1.21) conditions. No significant main effect was found for decision aid type, $F(2, 18) = 0.79$, $p = .47$.

Results: handling of arrival aircraft

Participants' ability to issue instructions to arrival aircraft was measured by their response times. Arrival reaction time was defined as the length of time (in seconds) an arrival aircraft spent waiting for instructions after entering the participant's area of control.

A repeated measures ANOVA was performed to test for differences in the mean Arrival Reaction Time between the Baseline (M = 21.23, SE = 0.24), TCV (M = 22.35, SE = 0.29), and Timeline (M = 28.74, SE = 0.36) conditions. No significant main effect was found for decision aid type, $F(2, 18) = 0.53$, $p = .60$.

Summary of results

The differences in efficiency of departure sequencing for the TCV and Baseline conditions were as expected; however, the lack of a significant difference between the Timeline and Baseline conditions was surprising. One possible explanation for this relates to some of the limitations of the optimization algorithm—because arrival aircraft were issued instructions by the participant, the algorithm could not predict arrival aircraft locations accurately enough to consider them. It was therefore up to the participant to ensure that following the algorithm-derived sequence did not present any separation issues between aircraft. This almost certainly led to some instances where a participant was not able to follow the exact sequence displayed on the Timeline decision aid. Though it is possible that this limitation could be eliminated with a sufficiently advanced algorithm, new human-automation-interaction issues would arise when the operator assumes that the automation is somewhat responsible for ensuring aircraft separation.

The lack of significant differences between all conditions for Departure Time Deviation is not particularly notable. With both decision aids being explicitly designed to improve the efficiency of departure sequencing, no additional support (beyond the interface design itself) was provided for sequencing departures according to scheduled departure time.

The lack of significant differences for Arrival Reaction Time does assuage some concern that the visual novelty of one or both decision aids might distract excessive amounts of attention from other areas of the display (specifically the areas related to arrival aircraft). It is possible, however, that increased workload in the Baseline condition may have helped negate any attentional differences between the display conditions. Further analysis pertaining to eye tracking data and mental workload is necessary to draw any conclusions pertaining to these issues.

Project summary

This study tested two distinct automated decision aids in a simulation of surface operations at Dallas-Fort Worth International Airport. Results indicate that the TCV decision aid, providing participants with perceptual representation of key system constraints, produced more efficient departure sequences than both the Baseline display and Timeline decision aid, which provided a "point solution" release sequence based on an optimization algorithm. These results indicate that the use of appropriate visualization-based decision aids can significantly improve performance in a complex dynamic environment, while allowing the human to retain full control over the decision-making process.

Before moving on to the next study illustrating our design approach, it is useful to consider these results at a more abstract level. One can clearly view the Timeline decision aid as a product of automation design consistent with the metaphor of displacement, as discussed previously in this chapter. The motivation behind its design is to essentially replace human with machine decision making, albeit that a provision is made for the human to override the prescriptions of the aiding system, that is, opening up the decision loop just enough for the human to occasionally intervene. The decision problem is framed in terms of a set of constraints on effective action selection, and an objective function that continually updates a mapping of system state to suggested actions with respect to these constraints and a set of numerical values attached to various decision options representing the goals of effective system control. In designing the TCV decision aid, one can see that we simply opened the decision loop even further, by creating visualizations that made abiding by these constraints fairly obvious to a trained controller, and allowing the human controller to apply his or her knowledge of the goals of the task to determine effective aircraft sequencing. The piloting study described in the following is consistent with this same general schema: empower the human operator with a continually updated visualization of the constraints that need to be obeyed in action selection (in the next case, flight safety envelopes), and open up the control loop so that a human (pilot) is able to determine effective control inputs based on his or her skill and knowledge, rather than by shutting the human out of the control loop with automated control law or laws, which would be the approach favored by a displacement metaphor for automation design.

Study 2: Visualizing flight envelope protection automation

In this study, we present a concept for coupling pilots and control automation along with display designs that integrate information from automated systems into the traditional primary flight display. Additionally, we provide a novel, prototype display located to the side of the primary flight display dedicated solely to exposing automation information. Our work centers around a dynamic Flight Envelope Protection (FEP) system augmented with logic for loss-of-control (LoC) prediction and prevention automation. Previous work addressing technological solutions to LoC prevention, especially in off-nominal conditions, appears in Belcastro and Jacobson (2010) and Belcastro (2011, 2012), using both visual and aural methods for notification and cueing, and adjustable autonomy (Kaber, 2012) in the way authority is partitioned between pilots and automation. Relatedly, Conner, Feyereisen, Morgan, and Bateman (2012) present an approach to cockpit display design using perceptual cueing to indicate corrective control actions that should be taken to avoid aircraft LoC.

At a qualitative level, the overarching concept is to reduce LoC through the novel use of a set of safety envelopes defined by a set of flight parameters and their associated boundaries or limits if LoC events are to be prevented. This vocabulary of envelopes and safety limits is extended to our display enhancements, and to a logic by which FEP automation selectively engages to compensate for combinations of pilot commands and environmental disturbances to maintain stability to prevent LoC events when detected. We believe that this form of joint, compensatory architecture for coupling humans and automation is much in the spirit of the "horse and rider" or "H-Metaphor" guideline for vehicle automation and interaction advocated by Flemisch et al., 2003, where the "horse" (automation) is capable of certain life preserving actions in the absence of the "rider" (human operator). In addition, the automation would necessarily be aware of how engaged the operator is at any given moment via operator manipulation of the flight controls and would react in a timely manner to hazards in the enviroment. Hence, the H-metaphor is a reasonable model for how automation should compensate appropriately for operator OOTL situations or when the operator is asserting direct commands to the system.

The iReCoVeR control architecture

Our approach toward reducing LoC events is part of a larger set of efforts to develop technologies to prevent incidents and accidents. The *Integrated Reconfigurable Controller for Vehicle Resilience* (iReCoVeR) is part of a research collaboration between researchers at the University of Illinois at Urbana-Champaign and the University of Connecticut, as an effort to develop technologies to prevent accidents and incidents resulting from LoC events in transport class aircraft. The iReCoVeR architecture is based on a combination of the Aircraft Integrated Resilient Safety Assurance & Failsafe Enhancement concept described in Belcastro's research (ibid.), technologies for fault-tolerant flight control (Hovakimyan & Cao, 2010; Hovakimyan, Cao, Kharisov, Xargay, & Gregory, 2011), fault detection and isolation (Lee, Snyder, & Hovakimyan, 2014), safe flight envelope estimation and detection (Tekles et al., 2014) and LoC prediction and prevention (Chongvisal et al., 2014).

Under this concept, the core subsystems of the iReCoVeR architecture work together to prevent the development of an LoC sequence by breaking the chain of events at its different stages. In addition, to ensure robust interaction between pilots and the iReCoVeR automation, the framework also incorporates the design and integration of both enhanced and novel prototype cockpit interfaces. These interfaces were conceived to make the behavior of the iReCoVeR automation transparent to the flight crew by providing timely and effective situation awareness, not solely about the aircraft, but also about the current and future operation of the developed automation. The iReCoVeR architecture in relation to the pilot, autopilot, and the automation-situation awareness interface displays are shown schematically in Figure 6.7.

From an aviation psychology perspective, the important items to note in Figure 6.7 include the central roles performed by flight envelope determination, prediction, and protection, the communication of this information to both an automated, resilient flight control system *and* to an automation "Situation Awareness Interface" informing the pilot about the dynamic activity of this automation. Note also a switch to enable pilot discretion over whether the aircraft will be under autopilot control or manual control at any given time. It is also important to note that by "manual" control in this context, all we mean is that the autopilot is disengaged: as will be discussed in detail, even while the pilot is in manual control, the FEP automation may yet actively compensate for pilot inputs to maintain adherence to safety envelopes, or it may be completely disengaged from the control loop by the pilot in which case it simply provides displayed information about the relation between aircraft state and these envelopes.

Flight envelope protection

In Tekles et al. (2014), the authors present the design and implementation of a dynamic FEP system, which is one of the core subsystems of the iReCoVeR architecture, as can be seen in Figure 6.7. The FEP system ensures

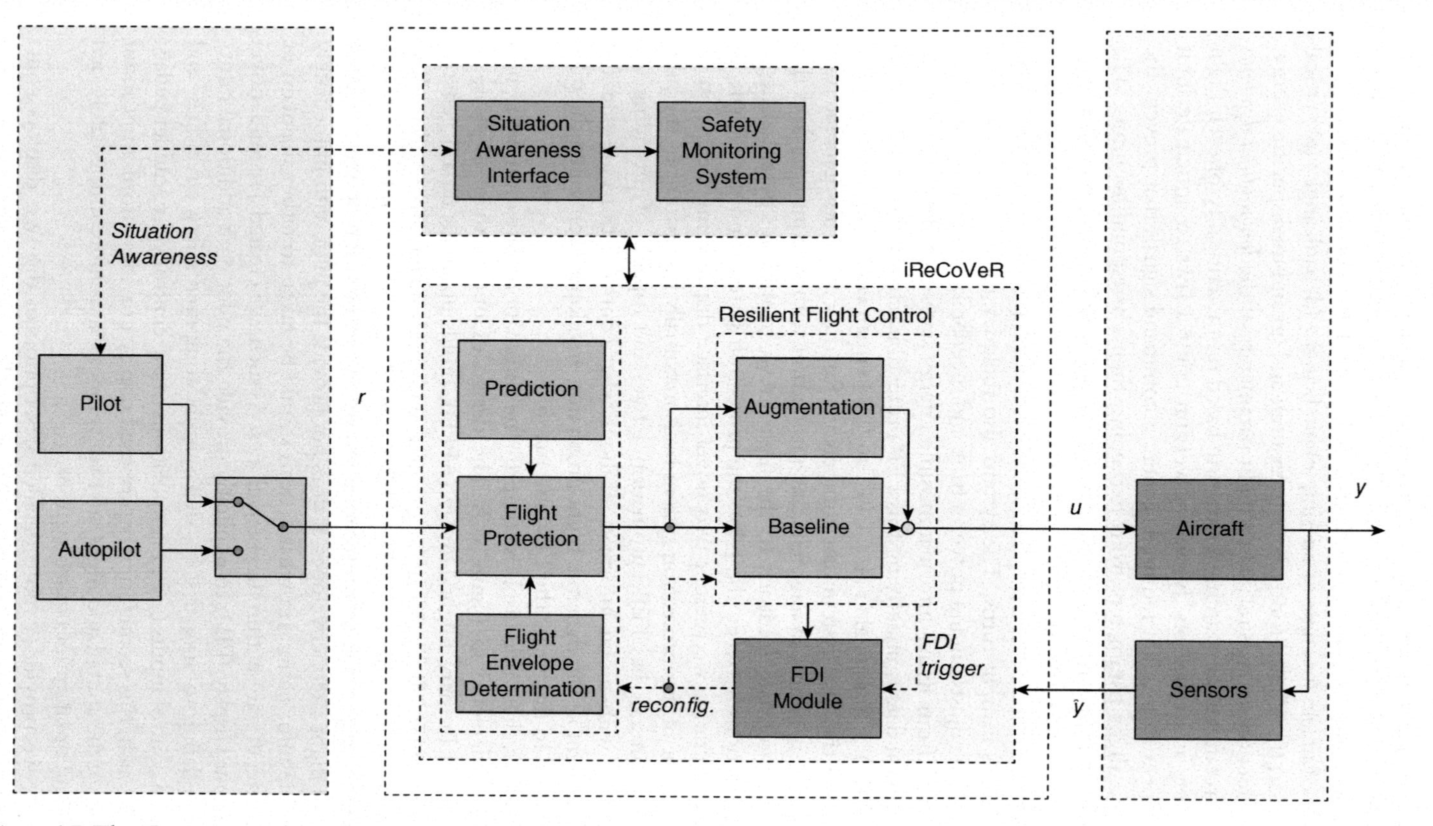

Figure 6.7 The iReCoVeR flight control architecture for LoC prevention.

Note. See Chongvisal et al. (2014) for additional details on the technical aspects of the overall control architecture. For additional information on the automation "Situation Awareness Interface," see the sections "Augmented primary flight display" and "Pilot input display" in this chapter.

that the aircraft remains within a predefined (dynamic) portion of the flight envelope that is determined to be safe. The protection scheme is based on a command-limiting architecture that is designed to protect excursions in: (a) angle of attack; (b) angle of sideslip; (c) pitch angle; (d) bank angle; (e) vertical load factor; (f) airspeed (or dynamic pressure); and (g) and total specific energy. The protection scheme related to flight parameters of the *longitudinal dynamics* of the aircraft has a hierarchical structure based on the criticality of each flight parameter, and relies on dynamic inversion control laws to anticipate limit exceedances and, when required, generate command signals that prevent the protected flight parameters from exceeding their corresponding limits. The *lateral-directional* protection scheme instead has a parallel architecture that relies on simple control laws designed to prevent excursions in bank angle and angle of sideslip. Moreover, a pitch-to-bank cross-feed prevents overspeed limit exceedances in tight turns. The system also includes a *total energy* protection scheme that adjusts engine power setting and speedbrake deflection to prevent the aircraft from flying into inadmissible energy states. Readers interested in the technical details underlying this FEP protection scheme should see Tekles et al. (2014), Chongvisal et al. (2014), and Ackerman et al. (2015).

Because the FEP system can compensate for the commands generated by the pilot, it is critical to augment this automation with cockpit interfaces that provide appropriate feedback to the flight crew, to mitigate potential negative interactions between the pilot, the protection system, and the aircraft. In the next sections, we present two novel interface displays intended to make the behavior of this protection system both maximally transparent to the pilot when the aircraft has FEP automation active, and capable of providing the flight crew with useful information about what control actions are safe and unsafe in light of the dynamic flight protection envelopes even when the pilot has selected to disengage the FEP automation.

Together, these two displays comprise the automation "Situation Awareness Interface" depicted in Figure 6.7. The design of one of these displays, an Augmented Primary Flight Display (APFD), builds upon a standard general-aviation primary flight display augmented with both quantitative and qualitative information provided by the FEP system. Importantly, while this display is intended to provide useful information on the relationship between aircraft state and flight safety envelopes, it does not provide the pilot with immediately actionable information on what actions should be taken to either maintain or return aircraft state within these envelopes. As such, we created a second, novel Pilot Input Display (PID) to explicitly provide the pilot with additional information relating to aircraft state, dynamically computed safety envelopes, and proximal control inputs (e.g., manipulations of the control yoke, rudder pedals, etc.). As such, this second, newly proposed display provides information about the state of the aircraft and flight safety envelopes that is more immediately tied to safe and unsafe pilot control actions.

Next, we provide a more detailed description of the two displays comprising our automation "Situation Awareness Interface," beginning with the APFD and followed by the PID.

Augmented primary flight display

The APFD, shown in Figure 6.8, is based on the standard primary flight display design, with the addition of three non-standard displays (Angle of Attack (AoA), Angle of Sideslip (AoSS), and load-factor). The elements of this augmented APFD are further enhanced by the addition of FEP-derived limits. Importantly, the framework for LoC used by our system uses sets of hard and soft limits to define envelopes around critical flight features (see Chongvisal et al. (2014) for details on how these predictive, "soft limits" are calculated by our LoC prediction algorithm, shown in Figure 6.7 as one component of the iReCoVeR architecture). A set of indicators (airspeed, pitch/roll, AoA, AoSS, and load-factor) are modified to display not only current status, but also their relationship current envelope position, both soft and hard. By providing salient cues concerning boundaries used by automation, pilots will become more aware of the reasons for automation engagement.

The general design for a limit indicator shows both hard and soft limits, the latter indicating impending exceedences of hard limits. For any given measurement, a yellow line is drawn parallel with the indicator movement. This line represents the range of values that are between the soft and hard limits. Moving into this region is an exceedance of the soft limits, and proper care should be taken that hard limits are not reached. The hard limit is marked at the end of the soft limit by a perpendicular yellow line. In the case of a soft limit excursion, this hard limit line turns red, drawing the pilot's attention. For example, on the AoA limit scale shown in Figure 6.8, the top of the vertical line represents the current AoA, while the horizontal line just above represents the soft limit. In Figure 6.8, the pilot is within the safety envelope defined for bank (e.g., the wings of the aircraft are level), but has pitched up to 25 degrees, thus exceeding the soft limit. The hard limit line in this instance changes from yellow to red, indicating a soft limit exceedance. We now describe the detailed aspects of this APFD display design.

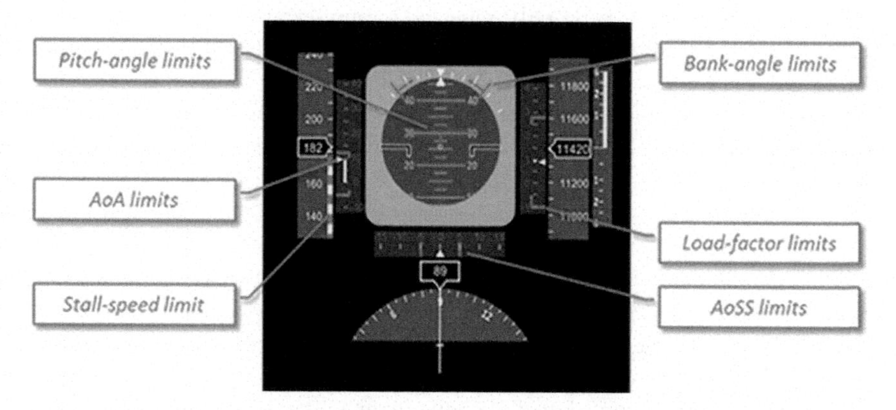

Figure 6.8 Primary flight display with FEP limit augmentations explicitly noted.

Airspeed tape

The airspeed indication is located to the left side of the instrument panel as part of the T-line concept (also referred to as the Basic-T); the "T" being formed by the command indicators of the airspeed, altitude, and neutral pitch bar aligned horizontally to each other. The indicator is of the vertical tape style with white numerals superimposed on a dark gray background. Major airspeed values indicated by tick marks at every 20 knots calibrated airspeed (KCAS), and unlabeled tick marks at every 10 KCAS. Current airspeed is indicated in a white bordered box marker and is centered vertically in the tape scale.

Additional symbology has been added to the airspeed indicator to provide the pilot with FEP-derived airspeed limit information. A pair of yellow bars is added to the airspeed tape to provide the pilot with an indication of the FEP upper airspeed limit. The soft limit is coded by a yellow bar that crosses the normal airspeed tick marks and extends to a large yellow tick mark that indicates the (dynamic) hard limit, nominally 350 KCAS. The lower limit for airspeed is indicated by a red and white tape superimposed on the airspeed tape, to maintain consistency with current practice for indication of critically low airspeed.

Angle of attack scale

Inset between the airspeed tape and the attitude indicator, the AoA indicator is a new display feature which represents the range of AoA values possible for this transport category model. Its location next to the airspeed tape facilitates a quick comparison during critical flight phases such as slow speed operations. As opposed to a current indication box marker, the AoA tape display utilizes a running bar scale with a chevron-shaped marker to indicate the current AoA value. Slightly less salient size and color differentiates this tape scale from the nearby airspeed tape. Major AoA increments are indicated by labeled tick marks at every 10 degrees, with unlabeled tick marks every 5 degrees in between. The full range is always displayed on the tape and is representative of simulation data.

FEP-derived limit symbols have also been added to the AoA scale. Upper and lower hard and soft limits are displayed to indicate the boundaries of the safe range of AoA. Soft limits are indicated by a yellow bar crossing the normal AoA tick marks and hard limits are indicated by yellow bars parallel to the normal tick marks.

Attitude indicator

Sizing of the attitude indicator display was modeled directly from a Rockwell Collins EFIS-700 electronic flight system attitude-direction-indicator unit. Display elements presented in the attitude indicator include a conventional pitch ladder with major labeled increments at every 10 degrees and minor increments every 2.5 degrees. Red chevrons are visible in the ladder at ±50 degrees and serve to advise the pilot as to the direction of neutral pitch during extreme attitude maneuvering. At the top of the attitude display is an angle of bank arc with major tick marks at 30 and 60 degrees and minor tick marks at every 10 degrees. By design, the angle of bank indicator does not serve as a sky pointer, but retains its aspect to the pitch indicator ladder during banked flight. In keeping with the standard coloring, the background area above zero degrees pitch is blue and below zero degrees is brown. Current pitch is indicated by a black pitch dot with miniature wings.

Four sets of soft and hard limit bars are added to the attitude indicator representing the FEP limits. For pitch limits, a vertical yellow bar crossing the pitch ladder indicates the soft limit and a horizontal bar parallel to the pitch ladder tick marks indicates the hard limit for both positive and negative pitch values. Bank angle soft limits are indicated by a yellow arc that crosses the angle of bank tick marks, while hard limits are indicated by a yellow bar parallel to the tick marks.

Altimeter tape

The altimeter indicator is of standard tape design and is located to the right of the attitude indicator. Major tick marks are labeled every 200 feet with minor unlabeled tick marks at 100 feet. Coloring is the same as for the airspeed indicator. Current altitude is indicated in a command marker box similar in appearance to that used for airspeed.

Vertical speed indicator

The vertical speed indicator is of standard design and located to the right of the altimeter indicator tape. Coloring is the same as for the airspeed indicator. The display has major tick marks appropriate to the range of the aircraft's performance. Major tick marks are labeled at 1,000, 2,000, and 6,000 feet-per-minute and minor tick marks at 500 feet-per-minute between zero and 2,000, and 2,000 feet-per-minute between 2,000 and 6,000. Current vertical speed is indicated by a white running bar that extends from zero. The zero point is level with neutral pitch, as well as the command box indicators for the airspeed and altimeter tapes.

Load factor scale

Between the attitude indicator and the altimeter tape is a new display to indicate load factor in the z-axis. Similar to the AoA scale, the load factor display utilizes a running bar scale with a chevron-shaped marker to indicate the current load factor value. Slightly less salient size and color differentiates this scale from the nearby altitude tape. The scale ranges from −1g to 4g with major load factor increments indicated by labeled tick marks at every 1g and unlabeled tick marks every 0.5g. The full range is always displayed on the tape and is representative of simulation data.

As with other protected flight parameters, upper and lower FEP limits are superimposed onto the load factor scale to provide the pilot with an indication of the aircraft's proximity to the envelope limits. Soft limits are indicated by a yellow bar crossing the normal load factor tick marks and the hard limits are indicated by a yellow bar parallel to the normal tick marks. The range of safe load factor values between the positive and negative hard limits depends on the location of the aircraft with respect to their envelope limits.

Angle of sideslip scale

The sideslip scale is a new display feature and represents the range of expected possible sideslip angles that can occur during flight for this aircraft model. The scale is located directly below the attitude indicator and uses the same layout and coloring as the AoA and load factor scales. The scale is zero-centered and ranges from −15 to +15 degrees with tick marks and labels every 5 degrees. Here positive values indicate right sideslip. The inclusion of this new display renders the standard split trapezoid sideslip indicator at the top of the attitude indicator redundant; therefore the split trapezoidal indicator is not displayed.

The FEP limits for this display element are indicated by yellow bars as a function of aircraft location with respect to hard and soft safety envelopes. Similar to the other limit indicators, soft limits are indicated by a yellow bar crossing the normal sideslip tick marks, and hard limits are indicated by yellow bars parallel to the normal tick marks.

Heading indicator

The heading indicator arc is located below the sideslip scale and completes the T-line arrangement of critical flight instrumentation. Presented as a partial arc, 110 degrees of heading are visible with a current heading command marker box at the top of the scale. A lubber line extends below the scale as in conventional horizontal situation indicators. Coloring of the scale is the same as the airspeed indicator. Tick marks indicate every 5 degrees of heading, with major tick marks every 10 degrees and labels at every 30 degrees.

Pilot input display

The APFD described above is primarily concerned with indicating the current status and the presence of limits in relation to dynamically computed FEP limits. However, the additional flight envelope augmentations on the APFD do not provide the pilot with directly actionable information, especially when the pilot has deactivated FEP automation and is flying the aircraft manually. When the automated FEP system is active, it directly limits pilot control input in response to potential envelope excursions. But when the FEP automation is deactivated, the APFD provides information to the pilot on the relationship of aircraft state, to safety envelopes, but *not* how to use cockpit controls to either maintain or to return this state within safety limits if and when they are exceeded. To communicate safety envelope information in a more direct relation to pilot control inputs, we designed a PID, located immediately to the right of the APFD, and depicted in two different modes in Figures 6.9a and b.

a

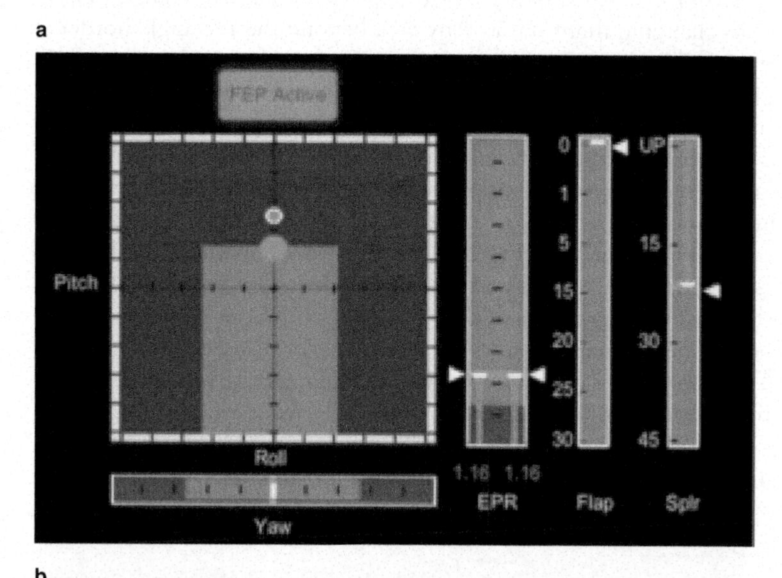

b

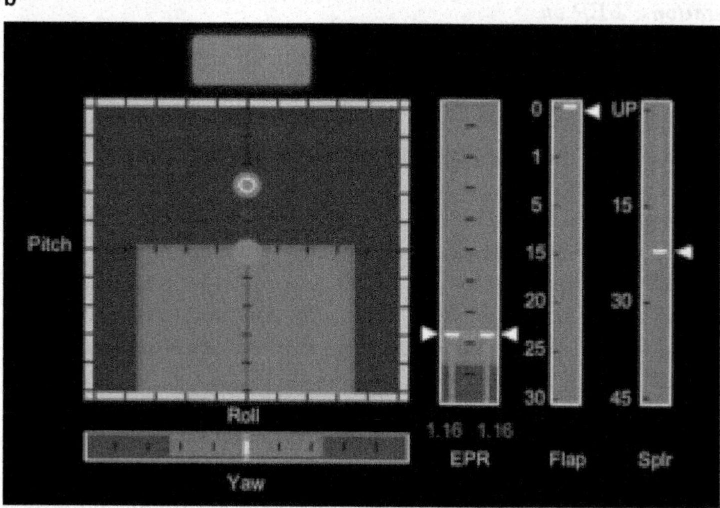

Figures 6.9 (a and b) The PID in FEP On mode and Active state (top), and in FEP Off mode (bottom).

There are two main elements to the PID: a square pitch/roll command box and a horizontal yaw command bar below the box. The box and the bar are marked by axis marks at regular intervals. The pitch/roll box and yaw bar depict the entire range of movement of the control yoke and rudder pedals respectively.

Within both display areas is a light gray rectangle bordered by yellow showing safe control inputs. Constraining control input to these rectangles guarantees hard FEP-derived limits are not exceeded. These rectangles move in response to changing flight status. Any area beyond the rectangle border is considered "unsafe" operation (e.g., entering the darker area of the rectangle in Figures 6.9a and b). Both displays are marked by two control input indicators. The first, a blue circle or bar outlined in white, represents the directed pilot input (shown in Figures 6.9a and b as the "bullseye" in the darker area of the rectangle). This always corresponds to the position of the yoke or rudder pedals. The second, also a circular, larger green indicator, represents the ideal FEP-derived control position. This marker will always remain inside the light gray box (shown in Figures 6.9a and b as the circle below the "bullseye" on the upper border of the light area of the rectangle). As seen in Figure 6.9a, the green (larger diameter) marker tracks the blue (smaller diameter) one indicating the safe position that is closest to the directed pilot input. When the pilot is directing input that is inside the light gray box, the two markers will overlap. At the top of the PID is an annunciator indicating the status of the FEP system (active or not). The behavior of this annunciator is dependent on the specifics of mode operation as described next.

Display operations: FEP on

One motivation for our design was the observation that during preliminary FEP flight trials, pilots were unaware of the impact FEP automation was having on aircraft control. When control input was modified during a difficult flight scenario, they perceived these modifications as a loss of (their) control. Despite the fact that the control system was actually maintaining aircraft stability effectively, pilots felt hindered by the system. As such, one motivation behind the design of the PID was to provide pilots with a more transparent window into the activity of FEP automation. This allows pilots to construct a more veridical model of overall system state, and readily accounts for situations where pilot control input requires active compensation from FEP-based automation.

While the FEP system is in the "On" mode, there are two states that are reflected on the FEP state annunciator. When the pilot is currently maintaining aircraft state and the directed input is within safety envelopes, the system is "Armed." This is noted by the FEP state annunciator being colored yellow and containing the text "FEP Armed." While in this mode, both blue and green indicator marks are overlapped, showing that the pilot is in complete control of the aircraft. The second possible state is "Active." This state occurs when the pilot directs a control position that is outside the displayed envelopes of safety.

At this point, the FEP modifies pilot input, directing the aircraft using a control position that is inside the safety envelopes. When the excursion takes place, the state annunciator changes from yellow to green, and displays the text "FEP Active." Additionally, the blue indicator will move beyond the yellow border, while the green indicator remains within the envelope. In this situation, the pilot is no longer in direct control of the aircraft—instead, the input displayed by the green indicator is being used.

Display operations: FEP off

While the primary configuration is intended to inform pilots of the system's state by making salient the impact of automation, it is possible to fly the aircraft without the use of the FEP system. Turning the FEP off allows the pilot to be in direct control of the aircraft, even in situations where the plane is in a potential LoC event. In this mode, the FEP continues to compute safety envelopes and ideal control positions, although it does not modify pilot input while continuing to display these envelopes. However, when the pilot directs the plane outside an envelope of safety (shown as a blue dot exceeding the yellow border, it becomes surrounded by a red halo, highlighting the excursion in the hopes of bringing this excursion to the pilot's attention). Additionally, the state annunciator turns red, communicating to the pilot that an envelope has been breached. The green marker remains inside the envelope of safety, marking an ideal or target position. This can be seen in Figure 6.9b, where the pilot has pitched up higher than is safely allowed.

Finally, we realize that the static views of the automation visualizations provided in Figures 6.8 and 6.9a and b, along with their narrative descriptions, are limited in their ability to communicate how the overall, integrated system functions in a dynamic, interactive context. As such, we encourage readers to additionally view a publicly available, five-minute video we have created to complement the information provided in this chapter (see FEP Demo, 2015, in the list of references for a link to this video).

Experimental evaluation

As part of our continuing research, we are currently performing experimental trials evaluating our design. We are performing pilot-in-the-loop testing at the flight simulator at the Illinois Simulator Laboratory. Our simulator is a Frasca 142 cockpit, with the primary flight display panel replaced with a digital display panel. Surrounding the cockpit are three projectors providing a 140 degree view of the outside world generated by X-Plane, while our physics model and FEP system are implemented in Matlab/Simulink. Our initial study participants are drawn from the Parkland College Institute of Aviation at the University of Illinois. We are recruiting both students and instructors, with the minimum requirement for participation being Private Pilot certification. In future research we intend to recruit active commercial transport pilots as participants.

Project summary

This work aims to create methods of visualizing information provided by advanced automation systems to pilots in salient ways that promote overall situational awareness. We believe certain failures are caused not by a failure in automation, or by a lack of pilot training or skill, but rather by insufficient communication between the two. We believe the insights gathered in this work may apply not only to the domain of aircraft automation and control, but to a variety of domains which necessitate automation systems and human operators working in conjunction. In particular, the language of safety envelopes seems particularly effective for a subset of these problems. Future work will be informed by the results of our simulator studies and a focus on developing generalizable methods for use in other application areas.

Conclusion

This purpose of this chapter has been to describe some design alternatives to default or even dominant approaches to automation design that adopt displacement or replacement approaches, in which automated systems are considered as fully substituting for functions or activities that have previously been performed by humans. The alternatives we have presented are based on the assumption that humans will continue to play vital roles in many technological systems or workplaces for the foreseeable future, regardless of the sophistication and intelligence of automated systems considered as their candidate replacements. When we are asked by many of our engineering colleagues why we are so interested in exploring these alternatives, the assumption seems to be that we are less optimistic than we should be about the tremendous pace of technological change to rapidly overcome any current limitations that continue, today at least, to give us pause about fully eliminating humans from the control loop.

In closing, we wish to stress that we are no less enthusiastic than our colleagues about engineering ingenuity—instead, we believe we are simply being realistic about the limits of engineering imagination. Any systems, such as those discussed in this chapter, that need to operate in open environments, that is, those not themselves fully created by human hand, will always have the potential to face unexpected situations requiring improvised solutions. We have presented human-automation interaction design approaches that certainly do not reject the many benefits automation brings. Yet we advocate augmenting these benefits with those that can be gained by using automation visualization to ensure that human operators are kept maximally informed about what their automated partners are doing and why they are doing so. We suggest that this approach is justified by the observation that, especially in unexpected and challenging situations, two heads are often better than one.

Authors' note

This research was supported by the National Aeronautics and Space Administration (Ames and Langley Research Centers) and by the National Science

Foundation, Cyber-Physical-Systems (CPS) program, Award #1330077. We also thank Bettina Beard, John Holbrook, Anna Trujillo, Christine Belcastro, Michael Byrne, Yijing Zhang, David Bauer, and Irene Gregory for useful discussions and guidance on the research reported here, and the volume editors for valuable feedback that substantially improved the presentation of this work. Portions of this chapter have been adapted from Kirlik et al. (2015), "Inverting the human/automation equation to support situation awareness and prevent loss of control," presented at the 2015 International Symposium on Aviation Psychology. Portions related to Study 1 have been adapted from Riddle et al. (2012), "A comparison of visualization and command-based decision aiding in a simulated aircraft departure sequencing task," in Proceedings of the Human Factors and Ergonomics Society 56th Annual Meeting, used with permission. Finally, portions related to Study 2 have been adapted from Ackerman et al. (2015), "Flight envelope information-augmented display for enhanced pilot situational awareness," AIAA Infotech @ Aerospace, Kissimmee, FL (AIAA 2015-1112).

References

Ackerman, K., Xargay, E., Talleur, D. A., Carbonari, R. S., Kirlik, A., Hovakimyan, N., Gegory, I. M., Belcastro, C. M., Trujillo, A., & Seefeldt, B. D. (2015). Flight envelope information-augmented display for enhanced pilot situational awareness. *AIAA Infotech @ Aerospace*. Kissimmee, FL. (AIAA 2015-1112).

Amelink, M. H. J., Mulder, M., Van Paassen, M. M., & Flach, J. M. (2005). Theoretical foundations for a total energy-based perspective flight-path display. *The International Journal of Aviation Psychology, 15*(3), 205–231.

Belcastro, C. M. (2011). Aircraft loss of control: Analysis and requirements for future safety-critical systems and their validation. *Control Conference (ASCC), 2011 8th Asian*, 399–406.

Belcastro, C. M. (2012). Loss of control prevention and recovery: Onboard guidance, control, and system technologies. *AIAA Guidance, Navigation, and Control Conference*. Minneapolis, MN. (AIAA-2012-4762).

Belcastro, C. M., & Jacobson, S. R. (2010). Future integrated systems concept for preventing loss-of-control accidents. *AIAA Guidance, Navigation, and Control Conference*. Toronto, Canada. (AIAA-2010-8142).

Bennett, K. B., & Flach, J. M. (2011). *Display and interface design: Subtle science, exact art*. Boca Raton, FL: CRC Press.

Billings, C. E., & Woods, D. D. (1994). Concerns about adaptive automation in aviation systems. In M. Mouloua & R. Parasuraman (eds.), *Human performance in automated systems: Current research and trends* (pp. 264–269). Hillsdale, NJ: Erlbaum.

Borst, C., Flach, J. M., & Ellerbroek, J. (2014). Beyond ecological interface design: Lessons from concerns and misconceptions. *IEEE: Systems, Man, and Cybernetics, 99*, 1–12.

Borst, C., Mulder, M., & Van Paassen, R. (2015). Delft ecological design, TU Delft. Retrieved from http://www.delftecologicaldesign.nl/research/ (accessed January 13, 2017).

Brinton, C., Lindsey, J., & Graham, M. (2010). The surface operations data analysis and adaptation tool: Innovations and applications. *Proceedings of the IEEE/AIAA 29th Digital Avionics Systems Conference*, (pp. 1.B.5.1–1.B.5.11). Salt Lake City, UT: IEEE.

Burns, C. M. (2013). Ecological interfaces. In J. D. Lee & A. Kirlik (eds.), *The Oxford handbook of cognitive engineering*. New York: Oxford University Press.

Burns, C. M., & Hajdukiewicz, J. (2004). *Ecological interface design*. Boca Raton, FL: CRC Press.

Casner, S. M., & Schooler, J. (2014). Thoughts in flight: Automation use and pilots' task-related and task-unrelated thought. *Human Factors, 56*(3), 433–442.

Chongvisal, N. T., Tekles, N., Xargay, E., Talleur, D. A., Kirlik, A., & Hovakimyan, N. (2014). Loss-of-control prediction and prevention for NASA's Transport Class Model. *AIAA Guidance, Navigation and Control Conference*. National Harbor, MD. (AIAA 2014-0784).

Conner, K. J., Feyereisen, J., Morgan, J. & Bateman, D. (2012). Cockpit displays and annunciation to help reduce loss of control (LOC) or lack of control (LAC) accident risks. *AIAA Guidance, Navigation and Control Conference*. Minneapolis, MN. (AIAA 2012-4763).

Degani, A., & Heymann, M. (2002). Formal verification of human-automation interaction. *Human Factors, 44*(1), 28–43.

Endsley, M. (1993). Situation awareness and workload: Flip sides of the same coin. *Proceedings of the Seventh International Symposium on Aviation Psychology* (pp. 906–911). Columbus, OH: Ohio State University, Department of Aviation.

Endsley, M. R., & Kiris, E. O. (1995). The out-of-the-loop performance problem and level of control in automation. *Human Factors, 37*(2), 381–394.

Erzberger, H. (2004). Transforming the NAS: The next generation air traffic control system. *24th International Congress of the Aeronautical Sciences*. Yokohama, Japan: International Congress of Aeronautical Sciences.

FEP Demo (2015). Retrieved from: https://www.youtube.com/watch?v=gLZpFfXwGVQ#t=282 (accessed January 13, 2017).

Flach, J. M. (2017). Supporting productive thinking: The semiotic context for cognitive systems engineering (CSE). *Applied Ergonomics, 59*(Part B), 612–624.

Flemisch, F. O., Adams, C. A., Conway, S. R., Goodrich, K. H., Palmer, M. T., & Schutte, P. C. (2003). *The H-metaphor as a guideline for vehicle automation and interaction*. Hampton, VA: NASA Langley Research Center. (NASA/TM-2003-212672).

Hadley, G. A., Prinzel, L. J., Freeman, F. G., & Mikulka, P. J. (1999). Behavioral, subjective and psychophysiological correlates of various schedules of short-cycle automation. In M. W. Scerbo & M. Mouloua (Eds.), *Automation technology & human performance* (pp. 139–143). Mahwah, NJ: Erlbaum.

Hoang, T., Jung, Y., Holbrook, J., & Malik, W. (2011). Tower controllers' assessment of the Spot And Runway Departure Advisor (SARDA) concept. *9th USA/Europe Air Traffic Management Research and Development Seminar*, June 14–17. Berlin, Germany: EUROCONTROL.

Hovakimyan, N., & Cao, C. (2010). *L_1 adaptive control theory*. Philadelphia, PA: Society for Industrial and Applied Mathematics.

Hovakimyan, N., Cao, C., Kharisov, E., Xargay, E. & Gregory, I. M. (2011). L_1 adaptive control for safety-critical systems. *IEEE Control Systems Magazine, 31*(5), 54–104.

Jamieson, G. A., & Vicente, K. J. (2005). Designing effective human-automation-plant interfaces: A control-theoretic perspective. *Human Factors, 47*(1), 12–34.

Jung, Y., Hoang, T., Montoya, J., Gupta, G., Malik, W., Tobias, L., & Wang, H. (2011). Performance evaluation of a surface traffic management tool for Dallas/Fort-Worth International Airport. *9th USA/Europe Air Traffic Management Research and Development Seminar*. Berlin, Germany: EUROCONTROL.

Kaber, D. B. (2012). Adaptive automation. In J. D. Lee & A. Kirlik (eds.), *The Oxford handbook of cognitive engineering* (pp. 594–609). New York: Oxford University Press.

Kirlik, A. (1995). Requirements for psychological models to support design: Toward ecological task analysis. In J. Flach, P. Hancock, J. Caird, & K. J. Vicente (eds.), *Global perspectives on the ecology of human-machine systems*. Hillsdale, NJ: Erlbaum.

Kirlik, A., Walker, N., Fisk, A., & Nagel, K. (1996). Supporting perception in the service of dynamic decision making. *Human Factors, 38*(2), 288–299.

Lafrance, A. (2015). The high stakes race to rid the world of human drivers. *The Atlantic Monthly*, December 1, 2015.

Lee, H., Snyder, S., & Hovakimyan, N. (2014). An adaptive unknown input observer for fault detection and isolation of aircraft actuator faults. *AIAA Guidance, Navigation and Control Conference*. National Harbor, MD. (AIAA-2014-0026).

Mulder, M. (2014). Ecological flight deck design—The world behind the glass. In M. A. Vidulich, P. S. Tsang, & J. M. Flach (eds.), *Advances in aviation psychology* (1st ed.) Burlington, NJ: Ashgate Publishing Ltd.

Norman, D. A. (1990). The "problem" with automation: Inappropriate feedback and interaction, not "over-automation". *Philosophical Transactions of the Royal Society B: Biological Sciences, 327*(1241), 585–593.

Parasuraman, R. (1997). Humans and automation: Use, misuse, disuse, abuse. *Human Factors, 39*(2), 230–253.

Parasuraman, R., Sheridan, T., & Wickens, C. (2000). A model for types and levels of human interaction with automation. *IEEE Transactions on Systems, Man, and Cybernetics– Part A: Systems and Humans, 30*(3), 286–297.

Parasuraman, R., & Wickens, C. (2008). Humans: Still vital after all these years of automation. *Human Factors, 50*(3), 511–520.

Rasmussen, J. (1986). *Information processing and human-machine interaction: An approach to cognitive engineering*. New York: North-Holland.

Sarter, N. B., & Woods, D. D. (1995). How in the world did we ever get into that mode? Mode error and awareness in supervisory control. *Human Factors, 37*(1), 5–19.

Seppelt, B. D., & Lee, J. D. (2007). Making adaptive cruise control (ACC) limits visible. *International Journal of Human-Computer Studies, 65*(3), 192–205.

Sheridan, T. (2002). *Humans and automation: Systems design and research issues*. New York: Wiley.

Smith, A., Evers, C., & Cassell, R. (1996). Evaluation of airport surface surveillance technologies: Radar. *Proceedings from CIE International Conference on Radar* (pp. 535–538). Beijing, China.

Tekles, N., Xargay, E., Choe, R., Hovakimyan, N., Gregory, I., & Holzapfel, F. (2014). Flight envelope protection for NASA's transport class model. *AIAA Guidance, Navigation, and Control Conference*. National Harbor, MD. (AIAA-2014-0269).

Vicente, K., & Rasmussen, J. (1992). Ecological interface design: Theoretical foundations. *IEEE Transactions on Systems, Man, and Cybernetics, 22*(4), 589–606.

Wickens, C. D. (2008). Multiple resources and mental workload. *Human Factors Golden Anniversary Special Issue, 3*, 449–455.

Wickens, C. D., & Hollands, J. G. (2000). *Engineering psychology and human performance* (3rd ed.). Upper Saddle River, NJ: Prentice-Hall.

Wickens, C. D., Mavor, A., Parasuraman, R., & McGee, J. (1998). *The future of air traffic control human operators and automation*. Washington, DC: National Academy Press.

7 Experimental evaluation of varying feedback of a cognitive agent system for UAV mission management

Sebastian Clauß, Elisabeth Denk, and Axel Schulte

Today, military Unmanned Aerial Vehicle (UAV) mission management (Clauß & Schulte, 2014; Theißing & Schulte, 2014) and human autonomy teaming in UAV contexts (Cummings, Bruni, Mercier, & Mitchell, 2009; Strenzke & Schulte, 2011) are high-priority foci for research. In modern UAV systems, conventional automation (such as auto-flight systems) relieves the operator of high-bandwidth sensor-motor tasks and improves precision and performance for mission execution. Instead of manual control, the operator controls the aircraft intermittently through automation. Therefore, a Human Supervisor (HS) more or less continually monitors the automation. For this type of control relationship, the term Human Supervisory Control (HSC) has been established (Sheridan, 1992).

In general, the HS performs five *supervisory functions*. The HS determines the current objective and explores a viable strategy to achieve it, incorporating the use of the given means (*plan*). The HS enters commands to the automation (*teach*) and monitors the automation to ensure proper execution (*monitor*). If necessary, the HS intervenes (*intervene*) and may eventually learn from experience to perform better next time (*learn*) (Sheridan, 1992). The cognitive capabilities of the HS allow the overall system to react to individual challenges in the environment and the status of the UAV system and thus enable the HS to compensate for unforeseen events.

During this process, the human operator has to process heterogeneous feedback information from varying automation functions which are highly sophisticated to accommodate for broader aspects of mission scenarios. While the automation was originally intended to support the operator, it can often increase cognitive demands instead due to the increasing complexity of the overall system to be monitored. In this context, Bainbridge (1983) describes two *Ironies of Automation*, the first being the shift of human errors from manual control to the design and implementation of automation functions. The second irony is *Clumsy Automation* (Wiener, 1988), which supports the operator in low-stress situations, but cannot provide support in highly intense situations. For the supervision of automation functions in manned flight, Billings (1997) described four *Costs of Automation*: *complexity*, *brittleness*, *opacity*, and *literalism* that may induce a new type of human error as a cause of complex automation.

To minimize similar problems in the guidance and mission management of UAVs, we propose a cognitive agent with extended feedback. This contribution describes the development and evaluation of a UAV application using such a cognitive agent onboard. The cognitive agent features a combination of mission-independent and mission-dependent information with proprietary perception results to support the human UAV operator. Modern UAVs feature automation functions that relieve the operator of highly complex control tasks. However, *Clumsy Automation* could induce additional workload and may result in human error. Advanced automation should be used for operator support and allow novel interaction modes as we know from in human-human delegation and cooperation. Therefore, we introduce a cognitive agent aboard a UAV supervising the onboard conventional automation, in order to reduce workload and enhance capabilities. In a simulator study, we examined the effects of two different types of feedback provided by the cognitive agent to the human operator. Results show that extended feedback by the agent positively affects the operator, especially when in-flight re-planning is required. Hence, the implementation of such feedback is advisable for task-based guidance of highly automated UAVs.

Agent supervisory control

To tackle some of these issues, advanced automation should be used for operator support and allow novel interaction modes as we know from in human-human delegation and cooperation. We added higher cognitive capabilities such as decision-making, problem solving, and planning aboard the unmanned aircraft. A cognitive agent, artificially implementing certain cognitive capabilities, is introduced onboard the UAV to manage and control installed conventional automation systems. In terms of *Cognitive Automation* (Onken & Schulte, 2010), the agent works within the supervision of the human operator as a Supporting Artificial Cognitive Unit and serves as a link between the human pilot's mission management and the mostly automated UAV navigation, guidance, and control layer.

Figure 7.1 shows the resulting work system featuring a cognitive agent as an agent supervisor controlling the automated UAV system. The human operator interacts with the single cognitive agent, rather than with a multitude of automation functions. Based on the mission goals and constraints given to the HS as directives, the human decides on how to employ the semi-autonomous UAV system most effectively. For this purpose, the human delegates objectives to the agent, whose cognitive capabilities allow it to derive action plans to achieve the objectives. The agent plans and coordinates the application of the underlying automation using its implemented knowledge to translate between mission criteria and automation-specific actions. But the semi-autonomous agent does not have the authority to modify or specify its own objectives (Onken & Schulte, 2010). The agent supervisor formulates discrete, parametric commands for conventional automation and monitors their execution. In analogy

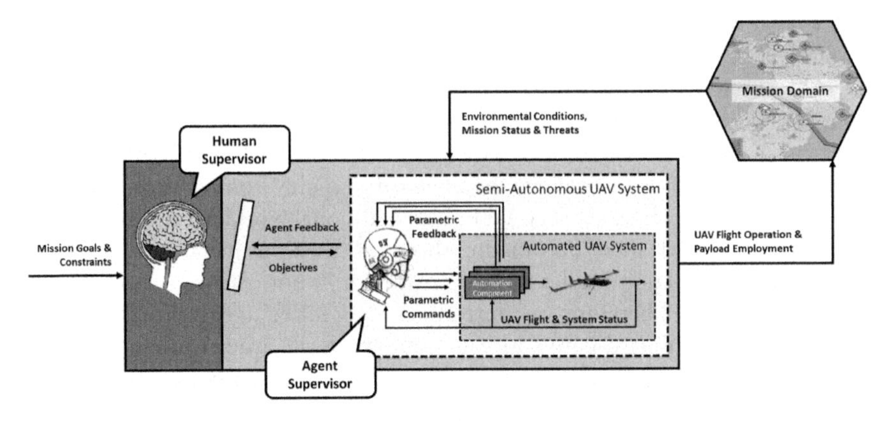

Figure 7.1 Work system of a semi-autonomous UAV featuring a cognitive agent as agent supervisor.

to the definition of HSC, the relationship between the cognitive agent, the conventional automation, and the UAV system may best be described by the term Agent Supervisory Control (ASC) (Clauß, Kriegel, & Schulte, 2013). The cognitive agent is acting like an *intelligent* subordinate to the human operator within the concept of HSC, using its cognitive capabilities to execute tasks previously allocated to humans and providing symbolic information as agent feedback to the human supervisor. The work system is embedded within a mission environment, in which it performs its flight operations and payload employment. The work system observes the environmental conditions including mission-relevant entities, the current mission status, and threats to the UAV system.

Figure 7.2 depicts the resulting guidance hierarchy of the UAV system, augmenting the automated system with an additional echelon, formed by the ASC loop. In this role, the cognitive agent combines two command relationships. On the one hand, it commands and monitors its underlying automation, and on the other, it serves as a subordinate to the HS.

Agent feedback within ASC

The human operator acting as a supervisor of the semi-autonomous UAV system requires information which allows monitoring of its performance and to intervene (re-plan) when necessary. In addition, the operator keeps control over whether to delegate a task to the agent or to perform it manually using the available automation systems. The operator's decision to delegate (teach) to the cognitive agent depends on the information conveyed by the system regarding the agent's capabilities and performance (Leana, 1986; Parasuraman & Riley, 1997). This feedback information is part of the human-agent interaction implemented into the system. For HSC of conventional automated

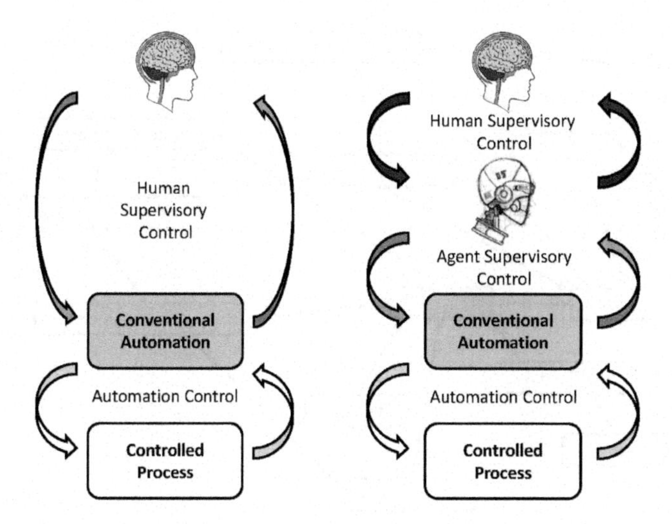

Figure 7.2 Guidance hierarchies within the concept of ASC.

systems, sophisticated concepts of interactions already exist. In the following, we examine an approach to a bidirectional information flow which allows both a calibrated delegation of tasks to the cognitive agent and adequate feedback to the operator.

Parasuraman and Riley (1997) present criteria for the task delegation to subordinate automation, which we mapped to a cognitive agent in Figure 7.3. The central criterion is *reliance*, representing the operator's willingness to delegate a task to the agent. The human reliance in the agent is directly influenced by the *confidence* in manual task execution, the current level of *fatigue*, the *perceived risk* associated with task failure, and the *trust in automation*, meaning the trust of the operator in the agent for satisfying task execution, based on the knowledge about automation states and behavior (*state learning*).

We identify three specific delegation criteria (*machine accuracy*, *trust in automation*, and *workload*) that are directly affected by agent behavior and, to a certain extent, controllable by agent and system design (cp. highlighted criteria in Figure 7.3). Machine accuracy describes the level of sufficiency with which the semi-autonomous work system, including the agent and the conventionally automated UAV functions, executes a delegated task. Task sufficiency, as perceived by the human operator, forms trust in the agent's capabilities to supervise underlying conventional automation. The agent's performance is rated with regard to directability, predictability of its actions, as well as task satisfaction (Klein, Woods, Bradshaw, Hoffman, & Feltovich, 2004). This behavioral information uses the human operator to predict future system behavior as well as the agent's capability to perform future tasks. The significance of operator trust in subordinate automation rises if the system complexity does not allow

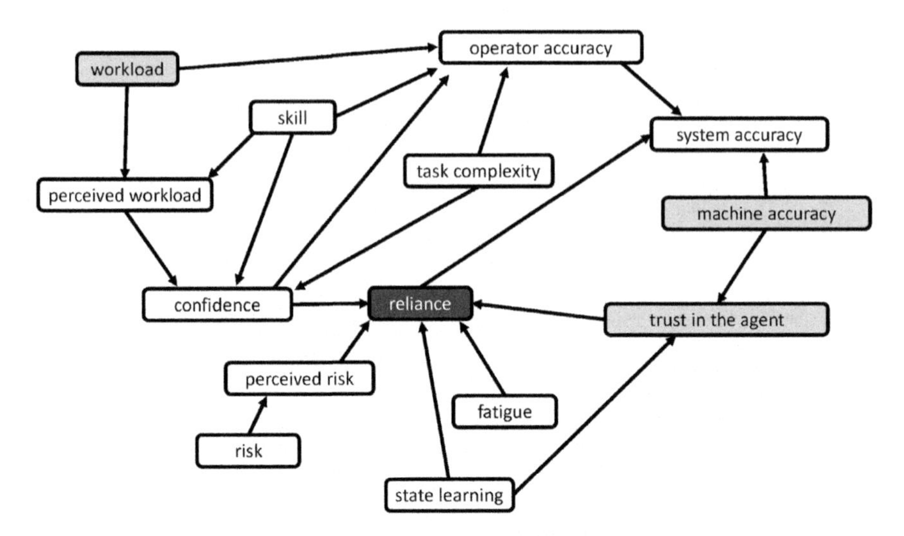

Figure 7.3 Criteria for human delegation to a cognitive agent (cf. Parasuraman & Riley, 1997).

Note. Criteria directly influenced by agent behavior are highlighted.

its full understanding and if strategies above predetermined, procedural behavior are employed (Lee & See, 2004). The interaction with automation during task execution directly influences the operator's sensory-motor and cognitive workload. In an ASC relationship, the agent behavior must aim at developing *calibrated trust* (Lee & See, 2004) referring to neither the over- nor the under-estimation of agent capabilities by the operator, allowing an appropriate decision on whether to delegate a task. The foundation for trust building regarding the agent's capabilities is the information conveyed from the agent to the operator. It may be categorized into three types (Lee & Moray, 1992): performance, process, and purpose. *Performance* information describes *what* and *how well* the agent is doing as well as the current agent status. The operator uses the performance history to derive the ability, reliability, and predictability of the agent. *Process* information gives the operator insight into *how* the agent works and by what means it is pursuing its delegated task. The operator derives structure from *process* information and gets an impression of the dependability and integrity of the agent. Finally, *purpose* information implies *why*, meaning for which purpose the agent was developed. When the purpose of the agent matches the tasks delegated by the operator, this information can enhance operator trust and extend dependency on the agent.

The form in which the agent conveys information to the operator also affects the development of trust. An established type of interaction, suitable for the ASC relationship and the current delegation (state), may be described by the term *etiquette* (Miller, 2002). Etiquette implies distinct knowledge about the context and the role of the agent within the work system and can enhance the effectiveness and safety of the overall system (Parasuraman & Miller, 2004). The adjustment of the interaction frequency and the temporal dependencies impact the overall benefit which can be achieved through ASC. Unintended alerts (Lees & Lee, 2007) may increase the operator's workload and lower his or her trust (*cry-wolf effect*). A lack of given information deprives the operator of the chance to intervene if required. The agent may initiate communication only in the case of events exceeding its knowledge and competence and when relevant to the mission guidance echelon, represented by the human operator. An example would be circumstances that prohibit the agent from accomplishing a delegated task within the boundary conditions specified by the human operator. The operator accomplishes the monitoring task by the assessment of the agent feedback. The agent itself monitors the heterogeneous conventional automation systems and creates a symbolic representation, which it communicates to the human. With regard to the operator's workload, the agent would effectively support the human if the cognitive task (i.e., interpreting the symbolic information) is less complex than the processes needed to monitor the heterogeneous conventional automation itself.

Feedback implementation within a task-based guidance approach

As a concept of delegation we chose a Task-Based UAV-Guidance (TBG) approach as already suggested earlier by Uhrmann & Schulte (2011) in a multi-UAV environment. Herein, the operator solely defines the objectives for the agent as commanded intents, instead of formulating step-by-step instructions for multiple automation components. Therefore, the operator defines what the semi-autonomous system shall accomplish, instead of providing how (meaning through what actions) this shall be achieved. Tasks might be military reconnaissance missions. The TBG design metaphor stems from an inter-human delegation relationship and relieves the supervisor of the tedious task of deriving automation action instructions from intentions (Clauß & Schulte, 2014). According to Theißing and Schulte (2013), tasks may be combined and scheduled within a task agenda. It includes specific tasks, intermediate steps, and their respective boundary conditions. It lies within the responsibility of the operator to ensure the agenda's achievability before assigning it to the cognitive agent. For the execution of a task agenda within the concept of ASC, we specify further the supervisory functions of the agent. According to the task-processing structure of Theißing, Kahn, and Schulte (2012), the step of planning can be divided into three consecutive processing steps.

Figure 7.4 shows the task agenda processing and the processing information within the plan-step of the agent for the example of a delegated reconnaissance task. In the first step, the agent supplements the delegated task agenda with missing (i.e., not entered by the human operator) agenda items logically required to reach the specified goal and respecting the specified task constraints. In the second step, the completed task agenda is reviewed by the agent according to overall mission constraints and rules of engagement. This way, the agent may detect violations stemming from conflicts in the task and mission constraints, as well as limitations of vehicle resources prohibiting the execution of the task agenda. The outcome of the review process is presented to the human operator for further consideration. In case of positive review feedback, the complete task agenda is displayed to the operator to allow an informed monitoring of the agent. If the review process is not passed successfully, the agent informs the operator and includes the reason for the negative outcome (i.e., the constraint conflict or the insufficiency of vehicle capabilities or resources). The final processing step is the decomposition of the task agenda, which results in a list of parameterized actions the agent has to implement during task execution by use of the underlying automation functions. For this purpose, the agent utilizes procedure definitions to translate certain tasks into control actions. During the implementation of the task (teach), the agent draws automation-specific action commands from the action list and monitors their execution.

In cases of external state changes resulting from environmental or system changes or events, the agent independently reevaluates its task agenda. If the current action list is still intact, the agent executes it as intended. Otherwise, parametric adjustments within the actions may be required to ensure a valid execution. If the adjustments impair the validity of the agent's task agenda, the agent re-runs the task processing in the same manner as for the initial task delegation. This way, the agent is able to derive an alternate task agenda (if possible) and implement it without human intervention, utilizing the strengths

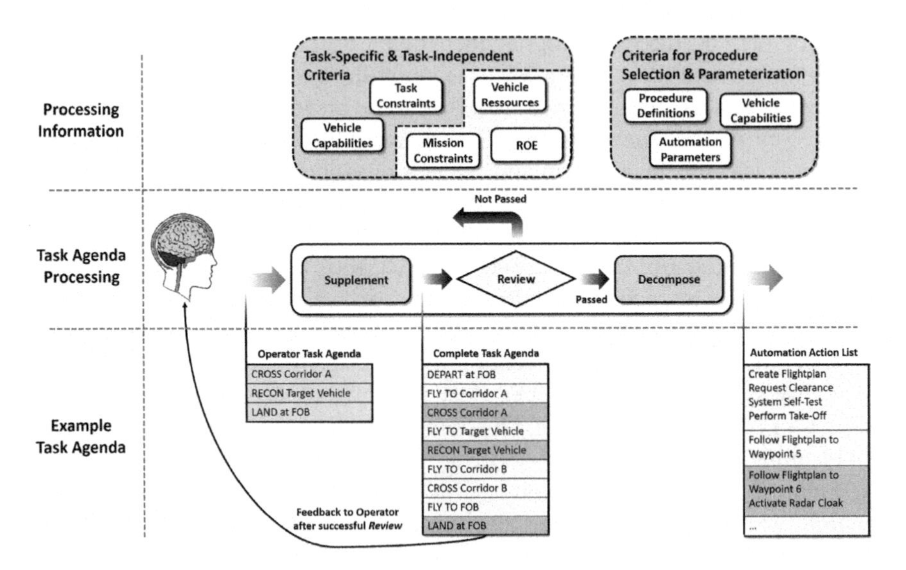

Figure 7.4 Task agenda processing by the cognitive agent using implemented information as part of its planning step.

of the goal-based TBG approach. As for the initial delegation, the complete task agenda is fed back to the human operator as process information. Only if the changes prohibit the agent from deriving a valid task agenda will the agent consult the human operator. In this case, the agent aborts the task execution and prompts further commands from the human operator.

Agent feedback intended to extend and support the human operator's monitoring task includes event-dependent and event-independent information. On standard events (e.g., task execution initiation and task completion), the agent informs the operator adequately, to enhance comprehensibility of the agent behavior. The state of the cognitive agent, its goals, the current task agenda, and the task execution status are also provided to the operator. The information is vital for the operator to assess the subordinate agent, its performance, and its ability regarding the task execution. The agent initiates communication with the operator if human clearance or assistance is required for task execution. In this case, the agent halts the execution of the task agenda until clearance is granted. The agent prompts the operator to revise the decision if clearance is denied. Also, through processing of automation information, the agent only allows the delegation of tasks, for which the UAV system currently provides execution capabilities. Finally, an environmental state representation is synchronized between the agent and the operator. This includes information about threats and tactical elements, representing objects in the physical world that the agent may consider during task execution.

The presented interaction concept is implemented using a graphical display as human-machine interface (HMI) for mission planning and execution—the Mission and Payload Control Station (MPCS) (Theißing & Schulte, 2014). For this purpose, the MPCS features a dynamic map showing the geographical outline of the mission area, tactical elements, and threats represented by NATO standardized icons (STANAG 2019) and a graphical representation of the UAV's flight plan including waypoints and route segments. Detailed information about tactical elements and the UAV are displayed by clicking onto their respective icons. A right click on tactical elements reveals a context menu with task alternatives that are available for the specific tactical element and in the current UAV state. The resulting task agenda is shown in the task agenda frame as an overlay in the upper left corner of the MPCS display. After its delegation to the agent, the task agenda is also depicted as arrows between tactical elements featuring specific task icons. The complete task agenda, returned by the cognitive agent after its review, is shown as *UAV Agent Plan* in the lower left corner of the MPCS display. The task execution status is also shown within the *UAV Agent Plan* to allow better agent monitoring. Agent-initiated communication and information are displayed as text boxes over the map area. Failure messages include the failure type and its reason with its respective tactical element or UAV capability.

The resulting HMI implementation is referred to as configuration B (Figure 7.5, right side). An additional benchmark configuration A (Figure 7.5, left side), equal in performance and structure of the cognitive agent, but with

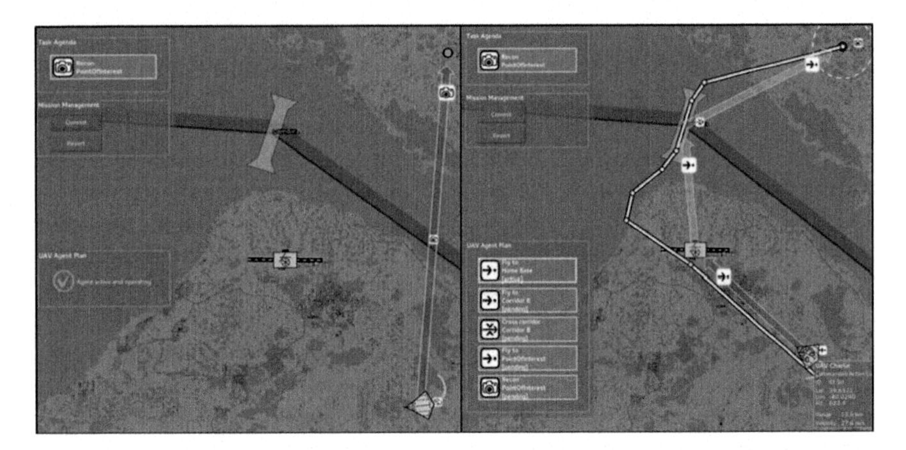

Figure 7.5 Moving map and planning display in MPCS (left: configuration A; right: configuration B).

solely graphical information about the position of the UAV and its current flight plan is used to evaluate the impact of the agent feedback design. In this configuration, the agent is still fully functional, but no information about its intent, resulting agent plan, or task execution is provided to the operator. Also, the human operator may delegate tasks independent of the current UAV capabilities. However, configuration A includes agent-initiated communication. While failure messages define their type, they do not specify their reason or the tactical element triggering the failure.

Experimental evaluation

The communication between the human operator and the automation on a symbolic level is an essential part of task-based guidance. We hypothesize that the feedback provided by the agent will affect the work result of the system, even if its decision-making and control functions remain unmodified. In an experimental campaign, we examine the impact of the agent's feedback behavior on planning, commanding, and re-planning the tasks of the human operator. The developed cognitive agent (Clauß & Schulte, 2014) with advanced feedback abilities is evaluated with respect to human performance, re-planning capabilities, and human trust in the agent. In our experiment, we compare our extended feedback system (configuration B) to a configuration with baseline feedback information (configuration A).

Research setup and configurations

The experimental design is a within-subject design with a secondary task (Borchers, 2014; Werner, 2014). The independent variable is the feedback configuration of the cognitive agent during mission execution (*configuration A* or *B*). For its evaluation, a comparative experiment was conducted in which two missions (*mission I* or *II*) are performed. Both missions have identical general layouts (see experimental procedure) which allow for comparison of their execution. The test group was divided into two groups, the first of which executed *mission I* in configuration A (limited feedback) and then *mission II* in configuration B (extended feedback). The second group executed *mission I* in configuration B and then *mission II* in configuration A. To eliminate sequence and spillover effects, we randomized the missions. The missions were stopped at predetermined positions during mission execution to obtain questionnaire answers from the participants.

The experimental hypothesis states that configuration B reduces workload, planning effort, and error rate, compared to configuration A. Also, configuration B leads to higher situation awareness, specific distribution of attention, trust in automation, visual perception, and acceptance compared to configuration A. In order to test the hypotheses, the constructs are operationalized using the following dependent measures.

As an objective performance measurement, the error rate of the human operator during delegation and manual re-planning was assessed. In addition, human-system interactions, such as the interaction time and the number of interactions were used as objective measures. Human interaction performance was assessed using an internal logging feature within the MPCS. In this context, we evaluated the interaction time and the number of mouse or touch-screen clicks during operator interactions. For planning and re-planning, the interaction time is defined as the duration between the initial human input and the final (correct) task delegation action, which we quantified by the number of manual actions (clicks) on the touch-display and by use of eye-tracking data taken during the monitoring task. The error rate is defined as the number of erroneously formulated task agendas prior to correct delegation.

As subjective performance measures, three questionnaires were presented during breaks within mission execution and after mission completion of each agent configuration. NASA-TLX (Hart & Staveland, 1988) was used to assess the subjective workload. The participants answered the six-scale questionnaire during predefined breaks during mission execution (subsequent to initial task delegation, during the monitoring phase, and after manual mission re-planning). SAGAT questionnaires (Endsley, 1988) were used to assess the situation awareness of the operators at specific points during the mission. For this purpose, the questionnaires were tuned to reflect specifics of the work situations and included questions to be answered both textually and graphically onto a map. The questionnaires were presented prior to the NASA-TLX during mission breaks in the monitoring phase and after manual mission re-planning. After mission completion in each agent configuration, the participants completed a custom questionnaire for the subjective evaluation of the acceptance of the system behavior and the trust in the cognitive agent's performance (Lee & Moray, 1992; Lee & See, 2004). The questionnaire consists of four dimensions (system interaction, system behavior, system information, and overall system), encompassing different statements, each rated by the experimental participants by assigning numbers representing their level of approval (1 = "not at all," to 7 = "fully applicable").

Participants and experimental conditions

The experimental test sample consisted of thirteen officers from the German Armed Forces. The participants were between 21 and 27 years of age (M_{age}=24.2) and were recruited from the University of the Bundeswehr Munich. Twelve male officers and one female officer participated in the experiment.

During the experiment, the operators controlled the UAV system from a Ground Control Station (GCS), using two vertically oriented multi-touch displays (Figure 7.6). Inputs could be made using touch-screen or mouse controls. Room dividers shielded the GCS to avoid visual distractions for the operators, while a headset dampened external sounds and facilitated communications. The lower screen featured the MPCS for TBG and automation monitoring using a map display of the mission area and a graphical representation of the tactical situation. The upper screen of the GCS featured a modular sensor interface showing live sensor-feed from the UAV. The sensor display was used to observe the surveillance area and to detect vehicle movements. The UAV's sensor was mounted face down below the UAV and offered a fixed image section for display. The GCS experimental setup included a faceLAB eye-tracking system, measuring the operators' focal point on the screens during the experiments. During the experiment, the MPCS facilitated the recording of frame positions and dimensions, which allowed the correlation of eye-tracking data and MPCS configuration during evaluation. The experimenter used an external workstation for the manipulation and control of the experiment's tactical situation and events (Borchers, 2014; Werner, 2014).

Experimental procedure

During the experiment, each participant successively performed *mission I* and *mission II*. Each experiment had a total duration of approximately three hours, including mission preparation, mission execution, and mission briefings and debriefings.

The experimental missions had an identical general layout with similarly complex mission scenarios, but different mission events and dynamic threats. The sample was split into two, where the first half performed *mission I* in system configuration A followed by *mission II* in system configuration B. The second half performed *mission I* in system configuration B followed by *mission II* in system configuration A. We randomized the assignment to the participant to minimize systematic effects.

The experimental test missions were as follows. An island, conquered by hostile forces, had to be retaken from an adjacent island, and therefore

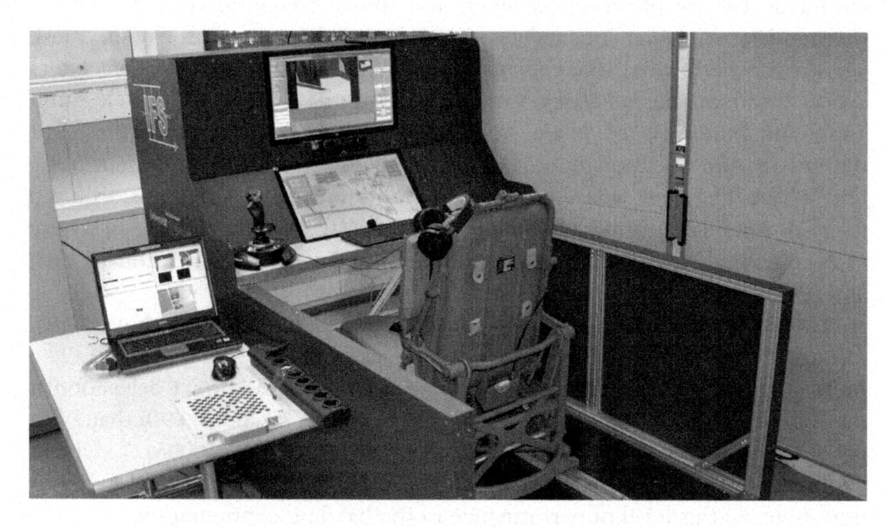

Figure 7.6 Experimental setup with operator workplace (right) and eye-tracking computer (left).

periodic UAV reconnaissance missions had to be performed by our test subjects. During the missions, own troops should be supported, hostile targets detected and identified, as well as areas scanned (reconnoitered). At the beginning of each mission, the UAV took off from its home base (indicated by the blue square below the UAV symbol) and crossed the Forward Line of Own Troops (FLOT, red line) through transit corridors (blue). Enemy Surface-to-Air Missile Sites (SAM-Sites), indicated in red, generally had to be avoided by the UAV. Reconnaissance targets were objects or areas (yellow). The main objective for *mission I* was to perform a detection task on two areas (A, B) and reconnaissance on two additional areas (C, D). The human operator was responsible for mission planning and execution. In *mission II*, a vehicle detection task had to be performed over area A and B and reconnaissance on area C. During the mission, the positions and ranges of hostile surface based air defense units were changed, as well as the availability of flight-corridors forcing the UAV to adapt its flight plan or its mission planning. The latter sometimes included the necessity for the operator to intervene and perform manual re-planning (if operator tasks contradicted the existing state).

Experimental results

We tested the experimental hypotheses by the application of a two-sided Wilcoxon signed-rank test using the SPSS software. For the exploratory data analyses, the significance level was set to 5 percent. For the data analyses, the mission was divided into three consecutive phases. The initial phase of delegation of tasks by the human operator (*delegation*), monitoring the agent's automatic task re-planning (*monitor*), and the intervention by the operator (*intervention*), in which manual re-planning and delegation of a revised task agenda was performed. For the analyses of differences in human-agent interaction based on the feedback variations, we only compared the delegation and manual intervention phase, since the automatic re-planning by the agent did not require operator input. For the subjective workload assessment using the NASA-TLX test, an additional baseline reference measure was acquired for each candidate during a familiarization mission. Measurement results were each compared between the agent configuration A with baseline feedback (FB) and agent configuration B with extended feedback.

The assessment of human error during manual intervention showed significant differences between agent configurations A and B (Figure 7.7). The number of erroneously formulated task agendas prior to correct delegation is significantly higher with baseline feedback (MW = 2.15; SD = 0.90) than with extended feedback (MW = 0.62; SD = 0.96; Z= −2.831; p = .005). Extended feedback resulted in zero errors for eight of the twelve participants, whereas none achieve this level of performance in the baseline configuration.

Figure 7.8 shows that there was no significant difference for the interaction time during initial delegation between configuration A (MW = 33.84; SD = 16.76) and configuration B (MW = 38.88; SD = 27.16; Z = −0.594; p = .588). During manual intervention, the extended feedback (MW = 50.03; SD = 44.54) significantly reduced interaction time compared to the baseline feedback in configuration A (MW = 112.57; SD = 58.38; Z = 2.432; p = .015).

The results of the interaction activity show similar results to interaction time (Figure 7.9). While no significant differences can be found between configuration

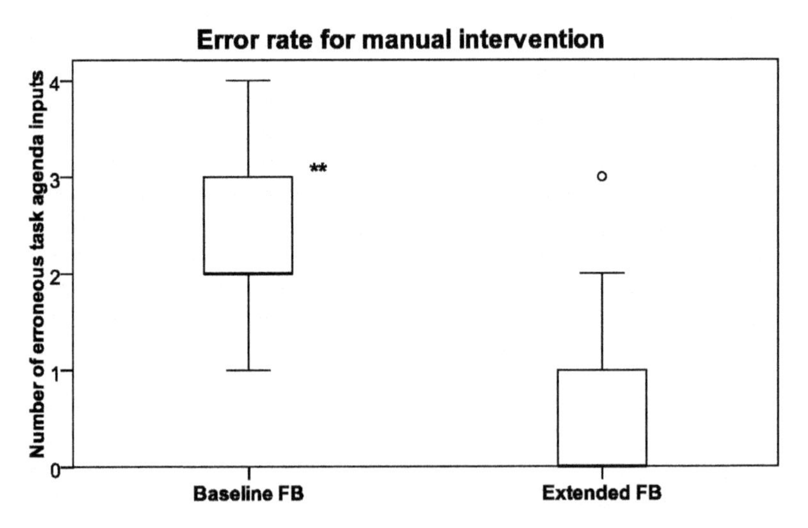

Figure 7.7 Error rate as number of erroneously formulated task agendas for manual *intervention* phase in agent configurations A and B.

** = p <0.01.

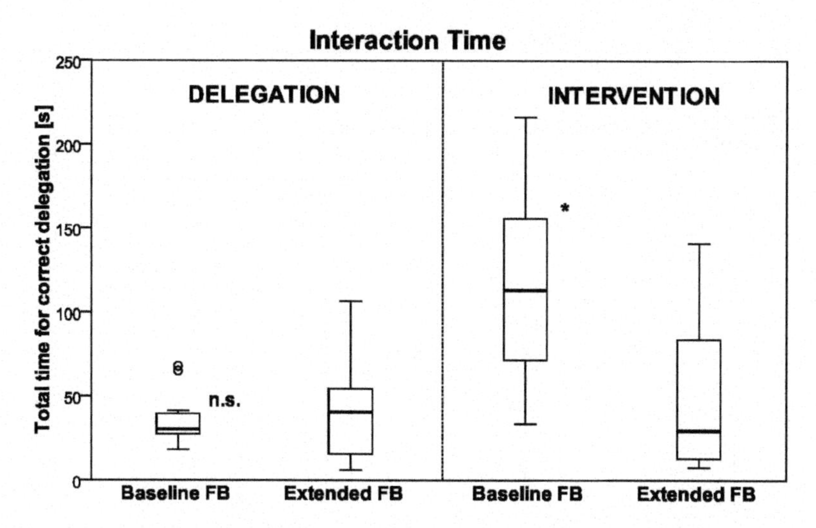

Figure 7.8 Interaction time as total time for correct task agenda delegation for the *delegation* and manual *intervention* phases in agent configurations A and B.

Note. n.s. = not significant, * = $p < 0.05$).

A (MW = 28.85; SD = 12.047) and configuration B (MW = 37.64; SD = 33.79; Z = −0.629; p = .554) during initial delegation, the extended feedback in configuration B (MW = 23.00; SD = 21.61) led to a significant reduction in the number of clicks compared to baseline feedback (MW = 68.46; SD = 51.31; Z = −3.059; p = .000).

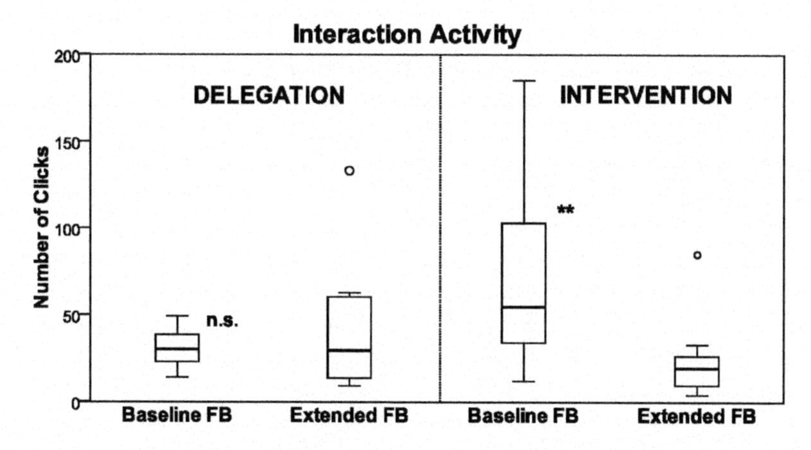

Figure 7.9 Number of interactions for correct task agenda delegation for the *delegation* and manual *intervention* phases in agent configurations A and B.

Note. n.s. = not significant, ** = $p < 0.01$.

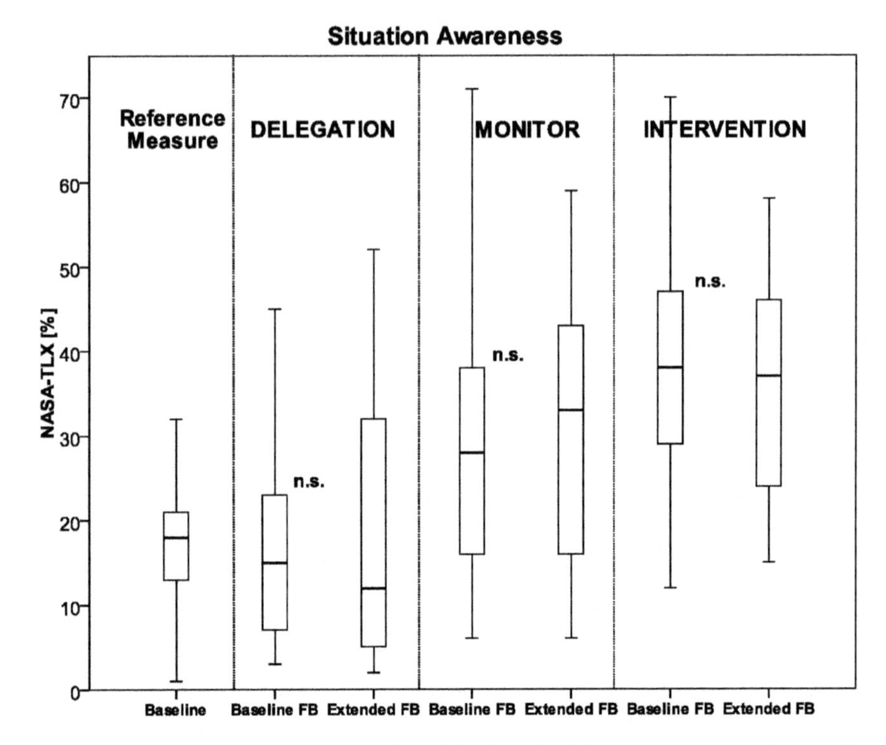

Figure 7.10 Subjective operator workload during the *delegation*, *monitor*, and manual *intervention* phases in agent configurations A and B, operationalized using NASA-TLX.

Note. n.s. = not significant.

The results of the NASA-TLX tests used to determine the subjective operator workload showed no significant differences between agent configurations in either of the three mission phases. The alteration of feedback does not significantly affect subjective workload during initial delegation (Z = −0.874; p = .414), the monitor phase (Z = −0.105; p = .946), or during manual intervention (Z = −0.559; p = .599) (Figure 7.10).

Situation awareness, determined by the SAGAT questionnaire during the monitor and the manual intervention phases, showed no impact of the agent configurations (Figure 7.11). During the monitor phase situation, awareness with baseline feedback (MW = 13.54; SD = 2.10) was not significantly different from extended feedback (MW = 12.61; SD = 1.93; Z= −1.325; p = .194). The same result was obtained for manual intervention where situation awareness in configuration A (MW = 11.14; SD = 1.98) was not significantly lower than in configuration B (MW = 12.47; SD = 2.76; Z = −1.577; p = .122).

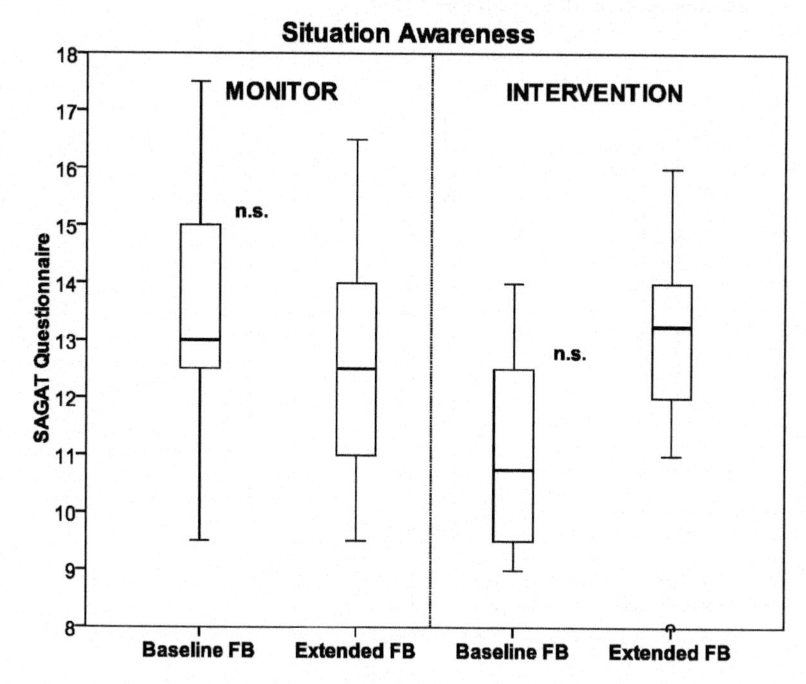

Figure 7.11 Situation awareness of the operators during the *monitor* and manual *intervention* phases in agent configurations A and B, operationalized using a SAGAT questionnaire.

Note. n.s. = not significant.

The evaluation of the post mission subjective questionnaire allows the depiction of system acceptance and trust (Figures 7.12 and 7.13). The operator's acceptance of the overall system performance is significantly higher with the extended feedback in configuration B (MW = 5.14; SD = 0.77) than with baseline feedback in configuration A (MW = 4.62; SD = 1.02; Z = −2.665; p = .005). The same is true for the operator's trust in the cognitive agent performance, where configuration A (MW = 4.44; SD = 1.08) led to lower trust than configuration B (MW = 5.06; SD = 0.77; Z = −2.518; p = .008).

Conclusions

This contribution describes an approach to cognitive agent feedback and its experimental evaluation concerning human delegation behavior and agent supervision. The developed cognitive agent is designed to support the human operator in the execution of supervisory control functions for a highly automated technical system. The agent is introduced in an additional ASC loop

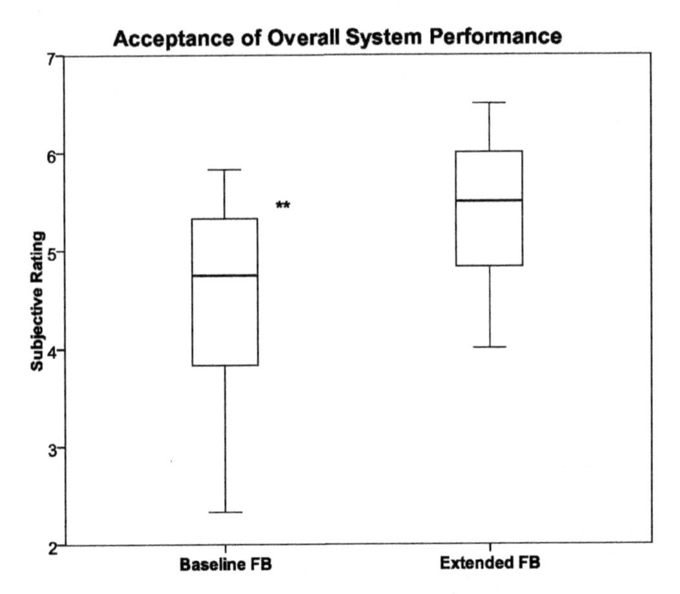

Figure 7.12 Acceptance of overall system performance in agent configurations A and B, subjectively rated via questionnaire.

Note. 1 = "not at all," 7 = "fully accepted"; ★★ = $p < 0.01$.

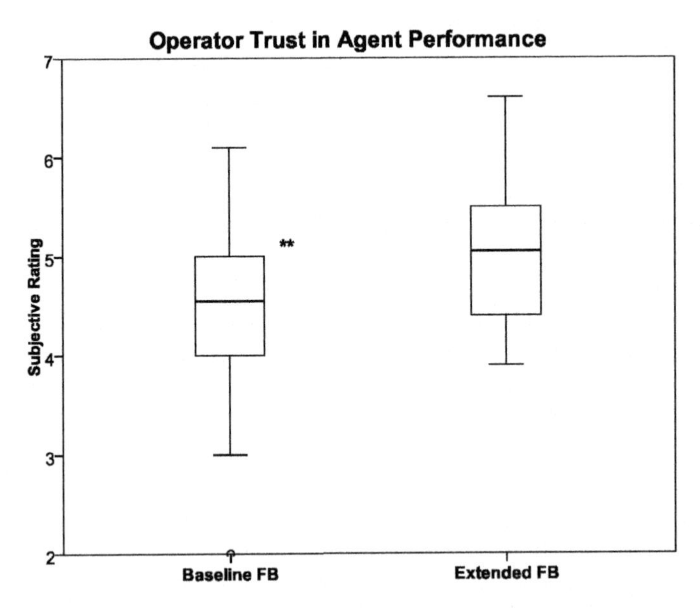

Figure 7.13 Operator trust in the cognitive agent's performance in agent configurations A and B, subjectively rated via questionnaire.

Note. 1 = "not at all," 7 = "fully trusted"; ★★ = $p < 0.01$.

raising the operator's guidance level to more abstract, symbolic mission guidance. A task-based delegation and processing approach is presented to enhance agent capabilities and to allow informed agent feedback to raise task execution flexibility and to simplify information processing for the human operator compared to conventional, sub-symbolic feedback. The implementation of adequate agent feedback in its role as a subordinate aims to create calibrated operator trust, a better understanding of agent behavior, and eventually the improvement of human-agent interaction.

Experimental results, presented in this contribution, confirm the claim for positive feedback of agent etiquette on human-agent relations, especially in the case of manual re-planning by the operator during mission execution. Meanwhile, the initial planning and the automatic re-planning phases of the mission remain unaffected. While the cognitive agent did not enhance the human operator's cognitive workload and situation awareness, the human interaction performance, trust in the agent, and the acceptance of the overall system performance were enhanced. Additional feedback information raised overall mission efficiency and lowered the operator's interaction and failure rate during mission re-planning. These results are consistent with a growing body of literature that emphasizes the important impact of interface representations for supporting effective collaboration between humans and autonomous systems (Parasuraman, Sheridan, & Wickens, 2000; Borst, Flach, & Ellerbroek, 2014; Chen et al., 2014). With increasingly autonomous components, success will often depend on increased attention to the design of interface representations in order to support rich communication between human and automated components to ensure the cooperation necessary for system success.

Future research may examine how erroneous agent behavior and feedback that disregards task objectives or constraints impact human trust and the inclination to delegate tasks to the agent. Existing research on conventional automation indicates an impact on both the number of tasks delegated and type, meaning the importance and criticality, of tasks delegated to the agent (Parasuraman & Miller, 2004). The presented results may also be extended regarding the impact of agent feedback on the ability of a single human operator to supervise multiple UAV. It might be reasonably assumed that existing results by Uhrmann and Schulte (2011) on single-operator multi-UAV operations can be improved with the presented human-agent interaction approach. Rudnick, Clauß, and Schulte (2014) have already employed the cognitive agent and the human-agent interaction approach on heterogeneous UAV platforms during a flight test campaign within a mission scenario similar to the one described in this contribution. Future flight testing may include human-machine experiments to verify the presented results outside of a controllable simulation environment.

Future operational use of UAV may benefit from ASC as an abstraction layer between the HS and the automation components aboard the aircraft.

In this sense, ASC may be used as a viable approach for modular UAV that implement specific types of airframes and onboard automation for different mission aspects, but which are all delegated by a single operator from inside a common GCS with a generic tasking and supervision interface. The characteristics of the agent feedback reflect the capabilities of the specific UAV configuration with respect to the work domain, and allow the operator to build and calibrate his or her trusts toward delegating tasks to the agent. This way, ASC can help to improve the system performance of UAV and to significantly reduce the familiarization period of human operators to unknown mission scenarios and UAV configurations.

References

Bainbridge, L. (1983). Ironies of automation. *Automatica, 19*(6), 755–779.

Billings, C. E. (1997). *Aviation automation: The search for a human-centered approach.* Mahwah, NJ: Lawrence Erlbaum Associates.

Borchers, A. (2014). *Konzeptionierung und Durchführung einer Versuchsreihe zur Evaluierung eines kognitiven Agenten in Bezug auf die Planung und Kommandierung von Aufklärungsmissionen [Design and execution of an experimental series to evaluate a cognitive agent with respect to planning and delegation in a reconnaissance mission].* Master's thesis. Neubiberg, Germany: University of the Bundeswehr Munich.

Borst, C., Flach, J. M., & Ellerbroek, J. (2014). Beyond ecological interface design: Lessons from concerns and misconceptions. *IEEE: Systems, man, and cybernetics, 99,* 1–12.

Chen, J., Procci, K., Boyce, M., Wright, J. Garcia, A., & Barnes, M. (2014). *Situation awareness-based agent transparency, ARL-TR-5905.* Aberdeen Proving Ground, MD: U.S. Army Research Laboratory (ARL).

Clauß, S., Kriegel, M., & Schulte, A. (2013). UAV capability management using agent supervisory control. *Proceedings of the AIAA Infotech@Aerospace Conference 2013.* Reston, VA: American Institute of Aeronautics and Astronautics (AIAA).

Clauß, S. & Schulte, A. (2014). Implications for operator interactions in an agent supervisory control relationship. *Proceedings of the IEEE International Conference on Unmanned Aircraft Systems* (pp. 703–714). New York: Institute of Electrical and Electronics Engineers (IEEE).

Cummings, M. L., Bruni, S., Mercier, S., & Mitchell, P. J. (2009). Automation architecture for single operator-multiple UAV command and control. *International Command and Control Journal, 1*(2), 1–24.

Endsley, M. R. (1988). Design and evaluation for situation awareness enhancement. *Proceedings of the Human Factors and Ergonomics Society Annual Meeting, 32*(2), 97–101. Santa Monica, CA: Human Factors and Ergonomics Society (HFES).

Hart, S. G., & Staveland, L. E. (1988). Development of NASA-TLX (Task Load Index): Results of empirical and theoretical research. In Hancock, P.A., & Meshkati, N. (eds.), *Human Mental Workload.* Amsterdam, Netherlands: North Holland Press.

Klein, G., Woods, D. D., Bradshaw, J. M., Hoffman, R. R., & Feltovich, P. J. (2004). Ten challenges for making automation a "team player" in joint human-agent activity. *IEEE Intelligent Systems, 19*(6), 91–95.

Leana, C. R. (1986). Predictors and consequences of delegation. *The Academy of Management Journal, 29*(4), 754–774.

Lee, J. D., & Moray, N. (1992). Trust, control strategies and allocation of function in human-machine systems. *Ergonomics, 35*(10), 1243–1270.

Lee, J. D., & See, K. A. (2004). Trust in automation: Designing for appropriate reliance. *Human Factors: The Journal of the Human Factors and Ergonomics Society, 46*(1), 50–80.

Lees M. N., & Lee J. D. (2007). The influence of distraction and driving context on driver response to imperfect collision warning systems. *Ergonomics, 50*(8), 1264–1286.

Miller, C. A. (2002). Definitions and dimensions of etiquette. *Proceedings of the AAAI Fall Symposium on Etiquette for Human-Computer Work.* Palo Alto, CA: Association for the Advancement of Artificial Intelligence (AAAI).

Onken, R., & Schulte, A. (2010). *System-ergonomic design of cognitive automation: Dual-mode cognitive design of vehicle guidance and control work systems.* Heidelberg, Germany: Springer.

Parasuraman, R., & Miller, C. A. (2004). Trust and etiquette in high-criticality automated systems. *Communications of the ACM, 47*(4), 51–55.

Parasuraman, R., & Riley, V. (1997). Humans and automation: Use, misuse, disuse, abuse. *Human Factors: The Journal of the Human Factors and Ergonomics Society, 39*(2), 230–253.

Parasuraman, R., Sheridan, T. B., & Wickens, C. D. (2000). A model for types and levels of human interaction with automation. *IEEE Transactions on systems, man, and cybernetics, 30*(3), 286–297.

Rudnick, G., Clauß, S., & Schulte, A. (2014). Flight testing of agent supervisory control on heterogeneous unmanned aerial system platforms. *IEEE/AIAA 33rd Digital Avionics Systems Conference (DASC 2014).* New York: Institute of Electrical and Electronics Engineers (IEEE).

Sheridan, T. B. (1992). *Telerobotics, automation, and human supervisory control.* Cambridge, MA: MIT Press.

Strenzke, R., & Schulte, A. (2011). Mixed-initiative multi-UAV mission planning by merging human and machine cognitive skills. In Harris, D. (ed.), *Engineering psychology and cognitive ergonomics* (pp. 608–617). Berlin, Germany: Springer.

Theißing, N., Kahn, G., & Schulte, A. (2012). Cognitive automation based guidance and operator assistance for semi-autonomous mission accomplishment of the UAV demonstrator SAGITTA. *61. Deutscher Luft- und Raumfahrtkongress 2012.* Bonn, Germany: Deutsche Gesellschaft für Luft- und Raumfahrt.

Theißing, N., & Schulte, A. (2013). Intent-based UAV mission management using an adaptive mixed-initiative operator assistance system. *AIAA Infotech@Aerospace Conference 2013.* Reston, VA: American Institute for Aeronautics and Astronautics (AIAA).

Theißing, N., & Schulte, A. (2014). Flight management assistance through cognitive automation adapting to the operator's state of mind. *Proceedings of the 31st Conference of the European Association for Aviation Psychology (EAAP).*

Uhrmann, J., & Schulte, A. (2011). Task-based guidance of multiple UAV using cognitive automation. *COGNITIVE 2011: The Third International Conference on Advanced Cognitive Technologies and Applications* (pp. 47–52). Wilmington, DE: International Academy, Research, and Industry Association.

Werner, J. (2014). *Konzeptionierung und Durchführung einer Versuchsreihe zur Evaluierung eines kognitiven Agenten in Bezug auf die Überwachung von Aufklärungsmissionen [Design and execution of a experimental series to evaluate a cognitive agent with respect to supervision in a reconnaissance mission]*. Master's thesis. Neubiberg, Germany: University of the Bundeswehr Munich.

Wiener, E. L. (1988). Cockpit automation. In Wiener, E. L., & Nagel, D. C. (eds.), *Human factors in aviation* (pp. 433–461). San Diego, CA: Academic.

8 Using simulation to evaluate air traffic controller acceptability of unmanned aircraft with detect-and-avoid technology

James R. Comstock, Jr., Rania W. Ghatas, Maria C. Consiglio, Michael J. Vincent, James P. Chamberlain, and Keith D. Hoffler

Unmanned Aircraft Systems (UAS) are no longer technological systems of the unforeseeable distant future, but rather of the present and near future. They are systems that are evolving quickly and will soon become commonplace in the National Airspace System (NAS). However, opening the NAS to civil UAS is a challenging task, a task that encompasses multiple safety issues which include detect-and-avoid implementations, self-separation procedures, and collision avoidance technologies to remain well clear of other aircraft. Routine access to the NAS will require UAS to have new equipage, standards, rules and regulations, and procedures, among others, in addition to many supporting research efforts to answer difficult questions concerning how these aircraft will operate in airspace with manned aircraft. As a result, the National Aeronautics and Space Administration (NASA) has established a multi-center "UAS Integration in the NAS" project to examine essential safety concerns, in collaboration with the Federal Aviation Authority (FAA) and industry, regarding the integration of UAS in the NAS. Among these guiding research efforts were the NASA Langley Research Center's Controller Acceptability Study (CAS) series, which looked at how Air Traffic Controllers would maintain traffic separation in busy airspace when some of the aircraft were UAS with detect-and-avoid equipment.

One of the major barriers to integrating UAS in the NAS is the requirement to see-and-avoid other aircraft per Title 14 of the United States Code of Federal Regulations (CFR), Parts 91.111 and 91.113 and other applicable regulations and accepted practices. In today's operations, pilots are required to follow right-of-way rules and remain well clear of other aircraft. There is also an obvious collision avoidance requirement. In all airspace classes, pilots are expected to comply with these see-and-avoid requirements while also complying with Air Traffic Control (ATC) instructions and clearances, or to negotiate changes to these instructions and/or clearances as necessary. See-and-avoid capable pilots are generally expected to maneuver and communicate in predictable ways and in a manner that preserves the safety, orderliness, and efficiency of the airspace system when operating in a positive control environment. UAS will likely be expected to operate in a similar manner, but with detect-and-avoid replacing the see-and-avoid capability of a manned aircraft. The acceptable design space and capabilities for detect-and-avoid systems in this environment are largely undefined. The controller-in-the-loop simulation experiments presented below sought to illuminate the detect-and-avoid design space for UAS operating in a positive control ATC environment.

Detect-and-avoid implementations also must be designed in a way that minimizes issuance of Resolution Advisories (RAs) by the Traffic Alert and Collision Avoidance System (TCAS) equipment on nearby aircraft. RAs are alerts with recommended vertical escape maneuvers to maintain or increase vertical separation with intruders that are predicted to be collision threats. RAs can be disruptive to the air traffic system and are a last resort maneuver when all other means of separation have failed. The detect-and-avoid concept evaluated in these studies was designed to detect encounter geometries that will cause an RA and provide guidance for action that may be taken early enough to avoid an RA.

In addition to avoiding the issuance of TCAS RAs, the detect-and-avoid algorithm should also be designed to prevent the issuance of controller traffic alerts, capture the attention of, or otherwise increase workload or create additional vectoring requirements for ATC. This work attempted to provide guidance for detect-and-avoid standards for well clear boundaries (following 14 CFR §91.113) that consider these ATC concerns.

Figure 8.1 shows a conceptual drawing of the different volumes and boundaries associated with remaining well clear of nearby traffic. In order to remain well clear of another aircraft, the detect-and-avoid well clear volume should be large enough to avoid (a) corrective RAs for TCAS equipped intruders; (b) safety concerns for controllers; and (c) undue concern for proximate see-and-avoid pilots. Determination of minimum and maximum operationally acceptable well clear volume sizes will inform the design space for the required detect-and-avoid surveillance range and accuracy. Guidance from the detect-and-avoid system will be provided to the UAS pilot to maintain positions outside the well clear boundary. Current standard NAS operations are the building blocks for which future UAS in the NAS operations will advance. In the two studies presented here, Air Traffic Controllers made acceptability ratings of varied Horizontal Miss Distances (HMDs) between UAS with detect-and-avoid technology and other traffic in the airspace. Acceptable distances are important as these values can be used to set requirements for detect-and-avoid algorithms for maintaining the UAS well clear of other traffic. In the concept under test in these experiments, for lateral maneuvers, headings that would result in a loss of well clear were displayed as amber bands on the heading scale of the navigation display. For the UAS pilot, the task would be to avoid flying headings where the amber bands appears. Details of the self-separation guidance shown to the UAS pilots to maintain well clear may be found in Chamberlain, Consiglio, Comstock, Ghatas, and Muñoz (2015). Research looking at the UAS pilot

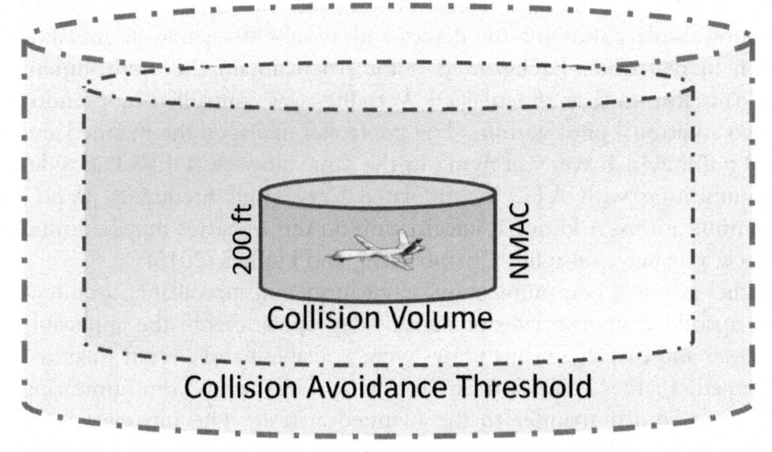

Figure 8.1 Notional depiction of detect-and-avoid well clear volume, collision avoidance threshold, and collision volume or near midair collision volume.

Note. Boundaries are notional and usually not cylindrical.

side of the detect-and-avoid ground control station display may be found in Comstock et al. (2016).

Developing the simulation

The scenarios developed for these studies simulated Dallas-Fort Worth (DFW) airport, ATC Sector DN/AR-7 South Flow, which is a portion of airspace delegated to DFW Terminal Radar Approach Control (TRACON) (D10). The majority of UAS traffic arrived or departed the Collin County Airport (FAA airport identifier: KTKI), about 28 nmi (nautical miles) northeast of DFW airport. The scenarios were designed and situated in this airspace so as to enable various encounter geometries between the UAS and intruder aircraft while manned aircraft traffic was handled in order to achieve realistic levels of workload for the controllers. Each test session was an hour in length and controllers handled an average of seventy-four aircraft in that time span with fourteen of those being UAS that would have encounters with other traffic requiring communications with ATC and in most cases a maneuver by the UAS. The level of traffic approximated what would be found in that sector on an average day. During development of the scenarios, an ATC subject matter expert on the research team visited DFW TRACON to insure the background traffic flow and rates used in the simulation were representative of actual operations.

For these studies there were a total of six test hours split over two days. Training and the first three hours of testing occurred on day one, with the remaining three hours of testing on day two. The UAS were controlled by two pseudo-pilots (working with research team), each having access to ground control station displays showing the detect-and-avoid self-separation guidance information in real time. Background traffic, to maintain the environment and workload close to that of actual DFW traffic, was controlled by pseudo-pilots at two additional pilot stations. The controller managed the manned and unmanned traffic which were all flying in the same airspace (DFW East-side) and communicating with ATC by simulated Very High Frequency (VHF) radio communications. Additional information on the scenarios may be found in Comstock, Ghatas, Consiglio, Chamberlain, and Hoffler (2015).

The studies assumed communication, navigation, and surveillance architectures and capabilities appropriate for current-day operations in the applicable airspace classes and that these capabilities were available to all aircraft (manned and unmanned) in the simulation environment. UAS were communicating with ATC in a similar manner to the manned aircraft. The intruders were Visual Flight Rules (VFR) traffic that were transponding but were not communicating with ATC. Additional Instrument Flight Rules and VFR manned aircraft traffic communicating with the subject controller included departures, arrivals, or overflights transiting the airspace. UAS command, control, and communication capability was assumed available between Unmanned Aircraft (UA) and their respective ground control station. The UA was assumed to be capable of receiving/transmitting voice communications to and from ATC facilities and proximate "party-line" aircraft via VHF radio in the same manner as manned aircraft in the same airspace, and of relaying these voice communications to/from the ground control station pilot via one or more UA-ground control station links. It was further assumed that, in addition to the relayed voice communications, the UA-ground control station link(s) carried all command/control data between the UAS and ground control station. This study assumed large size UAS (e.g., Predator or Global Hawk class). It was assumed that the UAS had surveillance sensors applicable to support detect-and-avoid and that the sensors were working and functioned without failures.

The displays for the UAS and manned aircraft control stations and the ATC displays were driven by modified versions of the Multi Aircraft Control System (MACS) software (Prevot, 2002), running on Windows-based computers. Modifications included incorporation of Stratway+ detect-and-avoid algorithms to drive navigation display "bands" which indicated a range of headings that would result in a loss of well clear with one or more traffic aircraft. These displays provided the information which guided UAS pilots to make requests for maneuvers in the encounter scenarios. Details of the appearance of the navigation display "bands" can be found in Chamberlain et al. (2015). Information on the detect-and-avoid algorithms may be found in Hagen, Butler, and Maddalon (2011), and Muñoz, Narkawicz, Chamberlain, Consiglio, and Upchurch (2014). The studies were run in a dedicated laboratory facility housed at Stinger Ghaffarian Technologies near the NASA Langley Research Center.

The flight paths and timing of the encounter aircraft (general aviation, transponding but not communicating with ATC) were scripted, meaning that the path and times were carefully designed to create the geometries and distances of the encounters under test. For the scripted traffic encounters used in both studies reported here, it is important to understand the meaning and differences of two terms: HMD, and closest point of approach. There was an adjustable parameter used in the detect-and-avoid well clear algorithms that yields a given HMD if the UAS pilot flies just at the edge of the guidance "bands." The UAS pilots, who were part of the research team, would fly at the edge of this guidance, yielding the desired HMD for each particular encounter (0.5, 1.0, 1.5, 2.0, 2.5, or 3.0 nmi in Study-1), so that controllers could evaluate that distance. Simulator data would yield the actual closest point of approach during the encounter as a check to insure that the desired HMD was obtained. The controller would look at the miss distance and geometry on the radar scope and evaluate the acceptability of that HMD as he or she saw it.

Closest point of approach had a second meaning in these studies as the scripted encounters had a designed or scripted closest point of approach if no pilot action took place. For example, both overtake and opposite-direction encounters had a scripted closest point of approach of zero, meaning that if no pilot action occurred then they would essentially hit each other. Because of the detect-and-avoid guidance, the UAS pilot would make a request to ATC to maneuver, and then pilot action occurred (a heading change) resulting in an actual closest point of approach generally very close to the HMD. In the case of crossing encounters, the closest point of approach and HMD relationship was more complex. If the scripted closest point of approach was the same as the desired HMD, then no pilot maneuver was required and the two aircraft would pass at the HMD (equal to scripted closest point of approach). In that case, no maneuver was required even if ATC notified the UAS pilot of the traffic. The UAS pilot would respond with "traffic detected" or say that the traffic was "no factor" based on the detect-and-avoid guidance. If the scripted closest point of approach was smaller than the HMD, then the detect-and-avoid guidance would result in the UAS pilot contacting ATC if time and radio frequency congestion permitted, and requesting a maneuver. The crossing scripted encounters were designed such that the UAS usually passed in front of the intruder to insure tight control over passing distances so that ATC could rate the acceptability of those distances. Real-world use of a detect-and-avoid system would yield maneuvers in front of or behind the intruding traffic depending on the geometry and other factors, as there would not be a scripted miss distance. The scripted intruder aircraft were flying under VFR and not talking to ATC, which would be legal in that part of the airspace, so all separation maneuvers were made by the UAS.

In Controller Acceptability Study-1, scripted closest point of approach crossing encounters of 1.5 nmi or less were designed to require no maneuver by the UAS to maintain the desired HMD. All opposite-direction (head-on) encounters and overtaking encounters did require communications with ATC and a heading change maneuver to achieve the desired HMD.

Study-1: determining acceptable miss distances

This study examined air traffic controller acceptability ratings of differing HMDs for encounters between UAS and manned aircraft transponding but not communicating with ATC. In this simulation of the DFW East-side airspace, the experiment tested HMDs that ranged from 0.5 nmi to 3.0 nmi. The controllers were tasked with rating these HMDs from "too small" to "too excessive" on a defined scale and whether these distances caused any disruptions to the controller and/or to the surrounding traffic flow.

The primary focus of the study was on ATC acceptability ratings of the differing HMDs used in the detect-and-avoid guidance to the UAS pilot for traffic encounters that were crossings, head-on, or overtakes. In addition to acceptability ratings, controller workload was also measured at five-minute intervals on a self-report subjective rating scale.

The aim of the study was to address, through data collection and analysis, the following research questions:

(a) Are detect-and-avoid maneuvers too small/too late, resulting in issuance of traffic safety alerts or air traffic controller perceptions of unsafe conditions?
(b) Are detect-and-avoid maneuvers too large (excessive well clear distances), resulting in behavior the air traffic controller would not expect and/or disruptions to traffic flow?
(c) Are there acceptable detect-and-avoid HMDs, in terms of ATC ratings, workload, and closest point of approach data that can be utilized in the development of detect-and-avoid algorithms?

For their acceptability ratings to be as valid as possible, fourteen recently retired ATC test subjects who had real-world experience controlling traffic on the East-side of the DFW airspace were utilized in this study. ATC experience among controller subjects ranged from 25.5 to 33 years with a mean experience 30.4 years. Controllers had an average of approximately 20.4 years of DFW experience in the TRACON Facility. Additionally, of that DFW experience, an average of 18.3 years' worth of experience was in the East-side sector of the DFW TRACON (D10) region. None of the controllers had experience with UAS operations in DFW airspace, and four of them were active instructors at the DFW training center.

With the aim of acquiring data on ATC acceptability ratings of differing spacing distances, the primary variable of interest was the HMD when traffic encounters occurred. A secondary variable of interest was the encounter geometry between the conflicting aircraft. Six HMD values were tested and included 0.5, 1.0, 1.5, 2.0, 2.5, and 3.0 nmi. These values were implemented in the detect-and-avoid algorithm, such that all distances were included in each

Table 8.1 Parameters of the primary and secondary variables of interest

Encounter geometry	*HMDs in separation algorithm (nmi)*					
	0.5	*1.0*	*1.5*	*2.0*	*2.5*	*3.0*
Opposite-direction	1 speed	1 speed	1 speed	1 speed	1 speed	1 speed
Overtake	1 speed	1 speed	1 speed	1 speed	1 speed	1 speed
Crossing	5 speeds	5 speeds	5 speeds	5 speeds	5 speeds	5 speeds

of the six one-hour test sessions. For example, this meant that an encounter where the encountering aircraft passed at 1.0 nmi could be immediately followed with an encounter where the aircraft passed at 3.0 nmi, such that all encounter HMD distances were subject to being rated by the controller during the test session.

Three encounter geometries were used, which included opposite-direction, overtake, and crossing. Table 8.1 shows the matrix of HMD by encounter geometry for the encounters. There were fourteen encounters per test hour, so three one-hour testing sessions were required to cover all cases, with an additional three one-hour test sessions on the second day permitting repeats of the entire matrix.

The following were the parameters used for the three encounter geometries. Opposite-direction: intruder (VFR manned) track at 180 degrees from ownship (UAS) +/− 15 degrees; crossing: intruder track at 90 degrees from ownship +/− 15 degrees; overtake: intruder track at 0 degrees from ownship +/− 15 degrees. The encounter speed differentials for crossing geometries were: 0, +/− 40, and +/− 80 knots. A single speed differential was used for opposite-direction and overtake encounters. All encounter geometries were without vertical separation, but did include climbing/descending trajectories (within 200 feet vertical altitude difference at encounter point).

Measuring performance

System performance metrics. Aircraft-to-aircraft separation distances obtained from the simulation software were used to confirm that the HMDs were correct and that the UAS pilots (on the research team) had kept the aircraft near the guidance provided during encounters. Additional variables expected to reflect a communications system under stress included operational errors and deviations, delays to aircraft in scenario, re-sequencing arrival aircraft, and voice communication errors, such as transposing information, call sign errors, and repeats. However, these additional metrics were either extremely rare or did not occur at all so were not available for statistical analyses.

Human operator performance metrics. Three different human operator performance metrics were recorded. After each traffic encounter occurred, subject controllers were asked by an ATC subject matter expert seated next to them

Table 8.2 HMD rating scale definitions

Rating	*Definition*
1	Much too close; unsafe or potentially so; cause or potential cause for issuance of a traffic alert
2	Somewhat close, some cause for concern
3	Neither unsafely close nor disruptively large, did not perceive the encounter to be an issue
4	Somewhat wide, a bit unexpected; might be disruptive or potentially disruptive in congested airspace and/or with high workload
5	Excessively wide, unexpected; disruptive or potentially disruptive in congested airspace and/or with high workload

Note. Fractional values, e.g., 1.5, were acceptable.

Table 8.3 Workload rating scale definitions

Rating	*Definition*
1	Minimal mental effort required
2	Low mental effort required
3	Moderate mental effort required
4	High mental effort required
5	Maximal mental effort required
6	Intense mental effort required

Note. Responses made on-screen at approximately five-minute intervals.

to rate the HMD based on a scale from 1 to 5. Table 8.2 shows the scale used and defines each of the HMD ratings. In addition, any brief comments by the controller with regard to that encounter were noted. At approximately five-minute intervals, workload was assessed using a modified Air Traffic Workload Input Technique (ATWIT) methodology (Stein, 1985). ATWIT was used to measure mental workload in "real time" by presenting auditory and visual cues that prompted the controller to press a number from 1 to 6 on the rating scale at about five-minute intervals to indicate the amount of mental workload experienced since the last rating. The response scale was built into the controller display software, and the scale definitions are presented in Table 8.3. In addition, an "end-of-hour questionnaire" was administered to each subject controller at the conclusion of each one-hour data collection session. Additional anecdotal information was obtained during a debriefing after completing the experimental testing.

Horizontal miss distances

Crossing encounters. Figure 8.2 illustrates the HMD ratings for the crossing encounter geometry. The graph shows that the controllers found the 1.0 and

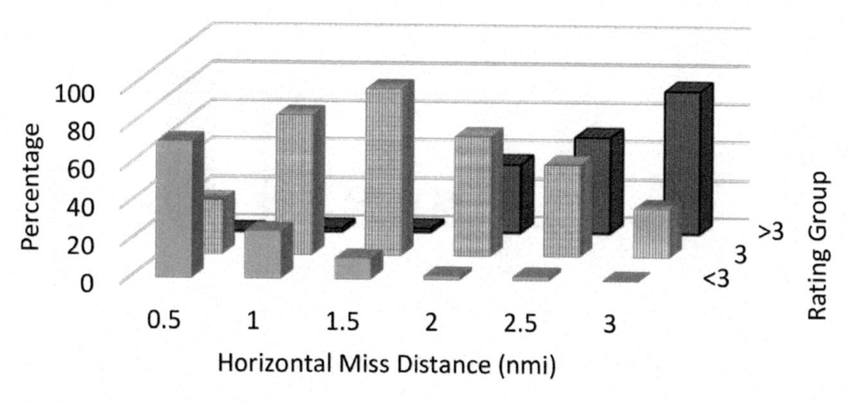

Figure 8.2 Subject controllers' ratings for HMD spacing parameters for the crossing encounter geometry.

Note. In these crossing encounters, the UA's speed was faster than the encounter aircraft.

1.5 nmi HMD spacing to be the most acceptable by giving a large majority of encounters with those specific spacing parameters a rating of 3, indicating that they were "neither unsafely close nor disruptively large," and "did not perceive the encounter to be an issue." HMDs of 2.5 nmi had comparable percentage ratings of 3 and more than 3. Furthermore, as was the case with the other two encounter geometries, HMDs with 3.0 nmi spacing parameters received a majority of ratings of more than 3, indicating that those encounters were either "somewhat wide," or "excessively wide," or "disruptive."

Opposite-direction encounters. Illustrated in Figure 8.3, the HMD ratings for the opposite-direction encounter geometry show that HMDs with a spacing parameter of 3.0 nmi were rated as being "somewhat wide" or

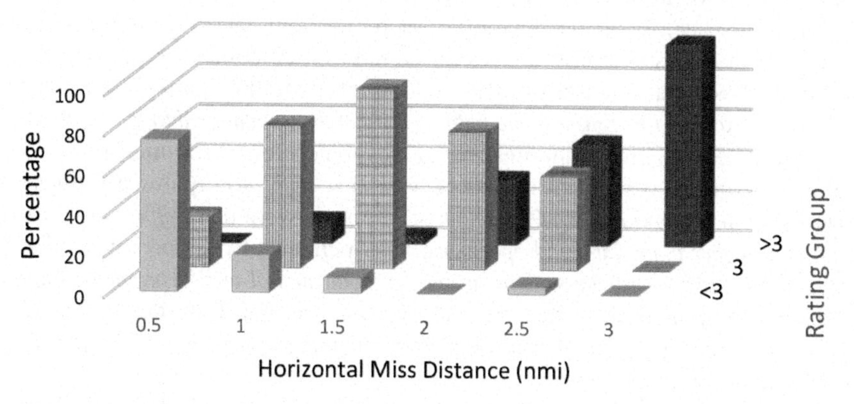

Figure 8.3 Subject controllers' ratings for HMD spacing parameters for the opposite-direction encounter geometry.

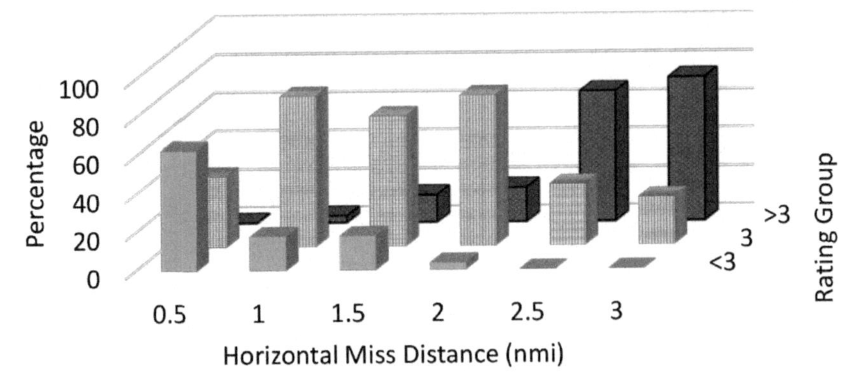

Figure 8.4 Subject controllers' ratings for HMD spacing parameters for the overtake encounter geometry.

"excessively wide." The graph also shows that the HMDs the controllers found to be acceptable were the ones in the 1.0 and 1.5 nmi range, with 80 percent of ratings suggesting 1.5 nmi being the more acceptable among the two.

Overtake encounters. Figure 8.4 illustrates the HMD ratings for the overtake encounter geometry. The graph shows that the highest percentages of "3" ratings were at the 1.0, 1.5, and 2.0 nmi HMD values. In addition, the graph also shows that a rating of more than 3 (wide or too wide) was given for HMDs of 2.5 or 3.0 nmi.

Realism of traffic density and workload ratings

Careful consideration was taken in the design and realism of the simulation environment with regard to traffic density. Research was conducted to find the optimal traffic density to achieve the aim of the study while maintaining as close to real-world densities as possible for a realistic simulation of the DFW East-side airspace. At the termination of each one-hour data collection run, an "end-of-hour questionnaire" was administered to each controller. Among the questions asked was one regarding the realism of the traffic density; controllers were asked to *rate the realism of the traffic density of the simulation during the preceding hour*. The following responses are for all subjects for all six one-hour data collection runs: 0 percent of responses were that *traffic density was significantly higher than in real operations*; 1.2 percent of responses were that *traffic density was somewhat higher than real-world operations*; 55.6 percent of responses were that *traffic density was about the same as would be found in real-world operations*; 42.9 percent of responses were that *traffic density was somewhat lower than real-world operations*; and 0 percent of responses were that *traffic density was significantly lower than in real-world operations*.

Table 8.4 shows the average workload ratings, captured at five-minute intervals using the modified ATWIT methodology, for all subjects and for

Table 8.4 Average ATWIT workload ratings

	Workload ratings (seconds during the hour)										
	300	*600*	*900*	*1200*	*1500*	*1800*	*2100*	*2400*	*2700*	*3000*	*3300*
Average rating	1.37	1.79	1.84	1.68	1.93	1.89	2.15	2.37	2.08	1.89	2.01

all data collection runs. These ratings were on a six-point scale (Table 8.3) so showed that despite the addition of UAS in the traffic mix, the workload reported by controllers remained low.

To summarize briefly, Study-1 employed a fairly high-fidelity simulation of the DFW East-side airspace and traffic density. The study focused on determining the effect of simulated detect-and-avoid-equipped UAS on ATC workload, as well as on the acceptability of maneuvers with differing HMD spacing parameters used in the detect-and-avoid algorithms. The results of the study confirmed a clear preference from the ATC ratings for the 1.0 to 1.5 nmi range. Based on their debriefing comments, the controllers found the detect-and-avoid self-separation concept as presented to them in this study to be completely viable. ATC workload ratings showed that the controllers rated workload on the low end of the rating scale even when handling about seventy-four aircraft per hour with UAS in the mix of traffic. Additional information on Study-1 may be found in Chamberlain et al. (2015), and Ghatas, Comstock, Consiglio, Chamberlain, and Hoffler, (2015).

Study-2: effects of winds and communications delays

The second study in the experiment series was based largely on the Study-1 experimental design and scenarios. The primary goals of this study were to address the impact of communication delays and wind conditions on the execution of ground control station detect-and-avoid and self-separation tasks and how acceptable the resulting maneuvers were to ATCs. The communications delays evaluated included four different ATC-pilot voice communication latencies or delays that might be expected in operations of UAS controlled by combinations of ground or satellite communications links. The latencies tested were 0, 400, 1200, and 1800 millisecond (ms) one-way communications delays. The communications delays were for the voice communications channel only and no delay was introduced for UAS control or position reporting.

One of the goals of the first study was to establish a generally acceptable HMD when there were encounters between detect-and-avoid-equipped UAS and transponder-equipped manned general aviation aircraft that were not communicating with ATC. The results of the earlier study indicated that HMDs of 1.0 to 1.5 nmi appeared to be acceptable for ATC when the traffic encounters are away from the airport vicinity. Based on those findings, in this study only HMDs of 0.5, 1.0, and 1.5 nmi were evaluated for encounters that were opposite-direction (head-on), overtakes (same direction with UAS faster), and crossings.

As in Study-1, fourteen encounters per hour were staged in the presence of moderate background traffic. In this study, controllers managed a mix of manned and detect-and-avoid-equipped UAS traffic and provided ratings on acceptability of HMDs when close traffic encounters occurred, and also provided workload ratings during the test conditions. Test conditions also included differing spacing parameters in the detect-and-avoid algorithms leading to different HMDs between the aircraft when they had close encounters, as well as varied voice communications delays and two levels of wind conditions.

As noted, a reduced set of HMDs for traffic encounters was tested in this study and included distances of 0.5, 1.0, and 1.5 nmi. In the first study, crossing scripted closest point of approaches of 1.5 nmi or less were designed to require no maneuver by the UAS to maintain the desired HMD. In this study, there were instances of crossing geometries of both 1.0 and 1.5 nmi that required maneuvers and concomitant communications with ATC. The minimum crossing geometry scripted closest point of approach was 0.5 nmi. All opposite-direction (head-on) encounters and overtaking encounters required communications with ATC and maneuvering to achieve the desired HMD.

The following research questions drove the experiment design:

(a) Given wind and communications delay conditions, were detect-and-avoid maneuvers too small/too late, resulting in issuance of traffic safety alerts or controller perceptions of unsafe conditions?
(b) Given wind and communications delay conditions, were detect-and-avoid maneuvers too large (excessive well clear distances), resulting in behavior the controller would not expect and/or disruptions to traffic flow?
(c) Given wind and communications delay conditions, were there acceptable, in terms of ATC ratings, workload, and closest point of approach data, detect-and-avoid miss distances that can be applied to the development of future detect-and-avoid algorithms?
(d) Do communications delays for the UAS in the airspace result in an impact on the ATCs' communications flow? Are the delays disruptive in terms of transmissions being "stepped-on" (simultaneous transmissions by several aircraft), and/or are additional repeats of information required with delays?

Seven recently retired ATCs with experience at the DFW East-side facility performed traffic separation tasks for the scenarios developed. Each of the controllers was the sole controller for the simulated DFW East-side environment for two days of testing. They were in communications with other controllers who handled handoffs to and from the subject controller's airspace. Most of the controllers were currently instructors in the training center at DFW. The majority of the controllers (five of seven) had participated in Study-1 conducted about four months earlier.

In addition to the HMD and encounter geometry conditions, additional variables of interest included two levels of wind (calm, about 7 knots; moderate, about 22 knots) and four levels of communications delay (0, 400, 1200, and 1800 ms one-way times). Communications with the UAS were handled in the same way as manned aircraft communications with the exception of the communications delays, which to the controller sounded as if the aircraft was taking a longer than normal time to respond to ATC.

The three HMD values were varied within the test hours such that a given encounter might follow an encounter with a different HMD. The wind conditions were varied by test hour with three hours having the lower wind value and three hours having the high wind value. Communications delay was also varied by test hour so that a given delay remained the same throughout the test hour. For a given test hour there were fourteen encounters consisting of ten crossing encounters (two at each speed differential), two opposite-direction encounters, and two overtake encounters. Each subject would experience six test hours across two days for a total of eighty-four encounters. Altogether, 588 encounter evaluations were made with 7 test subjects.

Measuring performance

Horizontal miss distance. After each traffic encounter, an ATC subject matter expert seated next to the controller subject asked: "How was the spacing of that last encounter?" or "How acceptable was the miss distance in the previous encounter?" Subjects had a copy of the information in Table 8.2 available to them during the test sessions.

Workload assessment. As in Study-1, about every five minutes during each hour-long test session, a workload rating was requested. This was done using the modified ATWIT method of workload assessment (Table 8.3). The subject clicked on their selection when prompted (i.e., an aurally presented—through headphones—"ding" occurred and the rating scale turned yellow).

System performance metrics. Data concerning the encounter aircraft separation distances were recorded throughout the period of the encounter, and included aircraft-to-aircraft separation distances and time to the closest point of approach. This information was obtained to confirm that the HMD was correct based on the guidance provided to the UAS pilot. For the communications time delay conditions, the communications system that permitted incorporating delays also recorded the push to talk status of all parties communicating so that "step-ons" (two stations transmitting at the same time) could be recorded.

Post-run questionnaires. After each one-hour test session, a questionnaire was administered to record ratings and comments on the preceding test session. Specific topics addressed included: 1—effects of communications delay; 2—realism of traffic density; 3—realism of workload; and 4—realism of communications rate.

Study-2: results

Figure 8.5 shows the mean HMD ratings by the controllers for each of the HMDs tested for the crossing traffic encounters using the scale in Table 8.2. The scripted closest point of approach was how close the two aircraft would pass if no maneuver was made. As noted earlier, if HMD was equal to scripted closest point of approach, then no maneuver would be called for by the self-separation algorithms, and no communications with ATC to request a maneuver were required. To see if the controller's rating was affected by whether the UAS had to contact ATC to request a maneuver to maintain the HMD, the encounter geometry was also set up such that the HMD was greater than the scripted closest point of approach for the 1.0 and 1.5 nmi HMDs. In Figure 8.5, the left-most bars at each HMD indicate situations in which the scripted closest point of approach was 0.5 nmi, which meant for the UAS pilot to obtain a 1.0 or 1.5 nmi HMD based on guidance provided, a maneuver was required. The diagonally shaded bars indicate a scripted closest point of approach of 1.0 nmi, so to obtain a 1.5 nmi HMD a maneuver was required. The cross-hatched bar (right-most bar at HMD of 1.5) was the case where both the scripted closest point of approach and HMD were the same so no maneuver was required. As can be seen from Figure 8.5, the acceptability ratings (bar heights) within a given HMD (1.0 or 2.0 nmi) stay the same regardless of whether a maneuver was required. This meant that controller ratings of HMD acceptability were not affected by whether communications with ATC and a maneuver were required by the UAS, which could have added to their workload.

Figure 8.6 shows the controller HMD rating data for crossing encounters and shows the highest percentages for a rating of 3 (*Neither unsafely close nor*

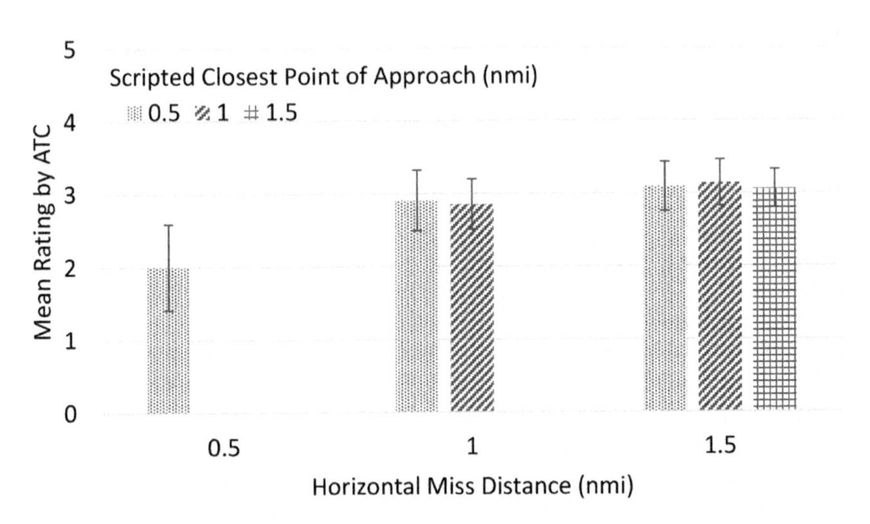

Figure 8.5 Mean HMD ratings by encounter distance for crossings.

Note. Error bars indicate +/− 1 standard deviation; rating scale definitions appear in Table 8.2.

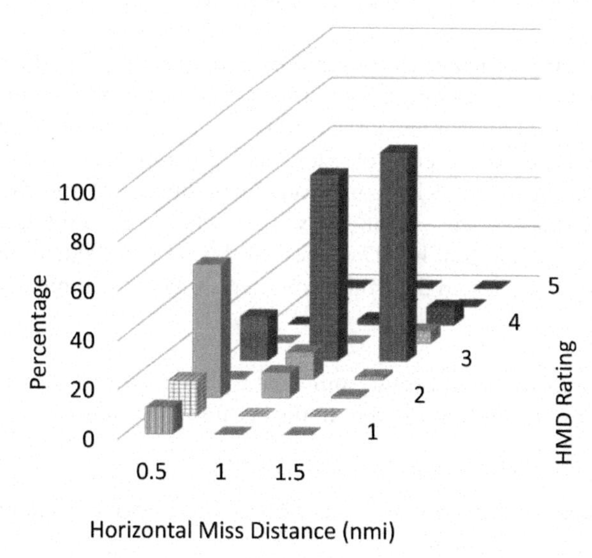

Figure 8.6 Ratings by HMD (crossing encounters).

disruptively large, did not perceive the encounter to be an issue), at the 1.0 and 1.5 nmi HMDs. Ratings shifted for the 0.5 nmi HMD, indicating greater concern for that miss distance. Figure 8.7 shows similar HMD rating data for the overtake and opposite-direction encounters, all of which required

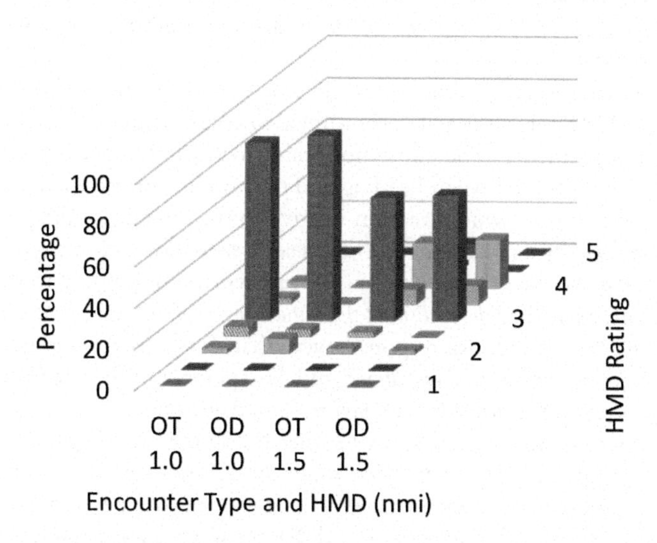

Figure 8.7 Ratings by HMD (overtake and opposite direction).
Note. 1 to 5 by 0.5; OT = overtake; OD = opposite-direction.

maneuvers to maintain detect-and-avoid prescribed HMDs, and communications with ATC.

Communications delays and wind. Communications delays of 0, 400, 1200, and 1800 ms were used for communications with the fourteen UAS per hour that had traffic encounters. The manned aircraft in the scenarios had no added communications delays. The occurrence of "step-ons" or simultaneous transmissions were found in all of the delay conditions (including zero) and their frequency or duration did not vary significantly across the delay conditions. Likewise, no differences in ratings of HMD or workload were noted due to the communications delays, but selected controller comments reflect the difficulties long delays introduce:

> The communications delays did cause some a/c to "step-on each other." This required extra transmissions to other traffic because they were blocked.
>
> The delay resulted in many repeats and was irritating.
>
> Repeats have a major impact on workload of ATC. In a busy environment you can't stand for a lot of them.
>
> Numerous repeats and step-ons! When in busy environments your transmissions need to flow and repeats/blocks only put you behind.

Also observed was a change in strategy by some controllers in the long delay scenarios to work manned, quicker-responding traffic first, then go to the UAS with their delayed responses. It should also be noted that the number of "step-ons" or simultaneous transmissions during the study may be less than what might occur in the real world, as pseudo-pilots and ground control station controllers cannot "step-on" themselves when controlling multiple aircraft in the simulation scenarios.

The "low" and "moderate" wind levels did not create any issues for the controllers. For the UAS pilots the self-separation algorithms handled the wind conditions with no problems. The most noticeable wind difference for the pilots was an expected offset between heading and track indicators on the navigation display, dependent on wind direction relative to the aircraft heading.

Realism of traffic density and workload. The research scenarios were designed to simulate traffic densities like those found in the real world. In response to the end of each hour question *Rate the realism of the traffic density of the simulation during the preceding hour*, 66.7 percent of responses were that *Traffic density was about the same as would be found in real-world operations*; and 31.0 percent of the responses were that *Traffic density was somewhat lower than real-world operations*.

The ATWIT workload ratings showed the following distribution of responses: 32.3 percent *Minimal mental effort required*; 42.9 percent *Low mental effort required*; 18.2 percent *Moderate mental effort required*; and 0.9 percent *High mental effort required*. Workload ratings did not differ across the two wind levels or four communications delay conditions.

Study 2: discussion

This study employed a simulation of the DFW East-side airspace with UAS operating in and out of Collin County airport northeast of DFW. The results confirmed the controller acceptability of 1.0 to 1.5 nmi HMDs found in the first study, even when maneuvers were required to maintain those miss distances, and winds were part of the scenarios. The 7 and 22 knot wind conditions tested were handled by the detect-and-avoid algorithms without issues and presented no problems for the controllers. Long voice communications delays between the UAS and ATC were identified as a problem in a high traffic density environment such as this, as reflected by controller comments.

The research questions presented earlier are repeated below with comments based on the results of the study.

(a) *Given wind and communications delay conditions, were detect-and-avoid self-separation maneuvers too small/too late, resulting in issuance of traffic safety alerts or controller perceptions of unsafe conditions?*

The acceptability ratings indicate that the 0.5 nmi HMD was judged to be "much too close" or "somewhat close, some cause for concern" for crossing encounters 78.6 percent of the time. This contrasts sharply with the crossing HMDs of 1.0 and 1.5 nmi where 75.2 percent and 84.6 percent respectively, rated these encounters acceptable ("Neither unsafely close nor disruptively large"). Encounter HMD ratings were not affected by the communications delay or wind conditions.

(b) *Given wind and communications delay conditions, were detect-and-avoid self-separation maneuvers too large (excessive well clear distances), resulting in behavior the controller would not expect and/or disruptions to traffic flow?*

Given that the largest HMD evaluated in this study was 1.5 nmi based on findings of the earlier study, it is not surprising that for crossing encounters at 1.5 nmi only 11.9 percent of the ratings indicated that the HMD was "Somewhat wide." While not a majority of the ratings, for the 1.5 nmi HMD for overtake encounters, 35.7 percent rated these on the "Somewhat wide" end of the scale. Similarly, for opposite-direction encounters at 1.5 nmi, 32.1 percent rated these on the "Somewhat wide" end of the scale. Based on discussions with the controllers, this is because it is easier to judge the HMD when aircraft are on parallel paths rather than crossing paths, especially when there are speed differences between the aircraft in the crossing scenarios.

(c) *Given wind and communications delay conditions, were there acceptable, in terms of ATC ratings, workload, and closest point of approach data, detect-and-avoid miss distances that can be applied to the development of future detect-and-avoid algorithms?*

The results of this study agreed with those of Study-1 and showed the greatest acceptability ratings for 1.0 and 1.5 nmi HMDs. For crossing maneuvers, an 84.6 percent HMD acceptability rating was found for the 1.5 nmi HMD. Also important to consider for future study is testing to insure the self-separation system maintains well clear boundaries so the collision avoidance systems, such as TCAS, are not activated by self-separation maneuvers. The detect-and-avoid algorithms handled the wind conditions without any problems for the UAS pilots (research staff) or for the controllers.

(d) *Do communications delays for the UAS in the airspace result in an impact on the ATCs' communications flow? Are the delays disruptive in terms of transmissions being "stepped-on" (simultaneous transmissions by several aircraft), and/or are additional repeats of information required with delays?*

The communication delays did not result in a greater number of "step-ons" or simultaneous transmissions and did not differentially affect controller workload ratings or HMD ratings, but based on debriefing comments, the longer delays would have been disruptive if traffic density had been higher. A number of the controllers changed their communications strategy in the longer delay conditions and worked the quick responding manned aircraft first, then going to the UAS with their delayed responses at less frequent intervals.

General discussion

When considering UAS, the initial thought might be that human error would be reduced if there is no human in the aircraft. However, removing the pilot moves the location where the human error takes place to the realms of the designer, developer, programmer, maintainer, as well as the remote pilot(s). Each of these entities can be a source of potential errors that can enter the system. At the same time, the final error-trap of a pilot on-board is lost, so that aspect of error reduction must be replaced by other means or technologies.

The present studies investigated through simulation methodologies aspects of the ATC controller's job in a future environment in which the traffic mix includes UAS with detect-and-avoid equipment. The information obtained from these studies is useful in the design of prototype detect-and-avoid systems that can serve to keep UAS separated from other traffic, both manned and unmanned. The research platform used was versatile and allowed control of many variables of interest including setting the delay of voice communications between ATC and the UAS, along with many other traditional flight operations parameters (e.g., number of aircraft in the scenario, wind).

Both studies presented here assumed perfect surveillance systems without positional error or uncertainty. Future studies are planned under the NASA UAS integration in the NAS project which incorporate sensor uncertainty and sensor effective range as variables of interest. Also of interest are simulations of failure modes, and especially from the ATC perspective, the maneuvers that a UAS would perform in a high traffic density environment if the communications or control link were lost. In the post-study debriefings, controllers expressed a strong need to know what to expect the aircraft to do in lost-link situations. The aim of this work is to provide useful information for guiding future rules and regulations applicable to flying UAS in the NAS.

Acknowledgements

The support of the many people involved in the conduct of these studies is gratefully appreciated. Thanks are extended to César Muñoz, Anthony Narkawicz, George Hagen, and Jason Upchurch of the Safety Critical Avionics Systems Branch at the NASA Langley Research Center for contributions to the development of the Stratway+ self-separation algorithms. Thanks are also extended to contributors participating through the NASA Langley Information Technology Enhanced Services contract, including: Pierre Beaudoin, Anna Dehaven, Steve Hylinski, Joel Ilboudo, Kristen Mark, Robb Myer, Gaurev Sharma, Jim Sturdy, Dimitrios Tsakpinis, and Paul Volk as well as other support and management personnel. Thanks are also extended to the tireless efforts of the persons staffing the background traffic pilot stations, the UAS pilot stations, and the ATC adjacent sector/tower consoles.

References

Chamberlain, J. P., Consiglio, M. C., Comstock, J. R., Jr., Ghatas, R. W., & Muñoz, C. (2015). *NASA controller acceptability study 1 (CAS-1) experiment description and initial observations*. NASA/TM-2015-218763, National Aeronautics and Space Administration, Langley Research Center, Hampton, VA.

Comstock, Jr., J. R., Ghatas, R. W., Consiglio, M. C., Chamberlain, J. P., & Hoffler, K. D., (2015). UAS air traffic controller acceptability study-2: Effects of communications delays and winds in simulation. In *Proceedings of the 18th International Symposium on Aviation Psychology* (pp. 318–323), Dayton, OH.

Comstock, Jr., J. R., Ghatas, R. W., Vincent, M. J., Consiglio, M. C., Muñoz, C., Chamberlain, J. P., . . . Arthur, K. E. (2016). *Unmanned aircraft systems human-in-the-loop controller and pilot acceptability study: collision avoidance, self-separation, and alerting times (CASSAT)*. NASA-TM-2016-219181, National Aeronautics and Space Administration, Langley Research Center, Hampton, VA.

Ghatas, R. W., Comstock, Jr., J. R., Consiglio, M. C., Chamberlain, J. P., & Hoffler, K. D. (2015). UAS in the NAS air traffic controller acceptability study-1: The effects of horizontal miss distances on simulated uas and manned aircraft encounters. In *Proceedings of the 18th International Symposium on Aviation Psychology* (pp. 324–329), Dayton, OH.

Hagen, G. E., Butler, R. W., & Maddalon, J. M. (2011). Stratway: A modular approach to strategic conflict resolution. *Proceedings of the 11th AIAA Aviation Technology, Integration, and Operations (ATIO) Conference*, September 20–22, 2011, Virginia Beach, VA.

Muñoz, C., Narkawicz, A., Chamberlain, J., Consiglio, M., & Upchurch, J. (2014). A family of well-clear boundary models for the integration of UAS in the NAS.

Proceedings of the 14th AIAA Aviation Technology, Integration, and Operations (ATIO) Conference, AIAA-2014-2412, Atlanta, GA.

Prevot, T. (2002). Exploring the many perspectives of distributed air traffic management: The multi aircraft control system MACS. *AAAI HCI-02 Proceedings*, pp. 149–154.

Stein, E. S. (1985). *Air traffic controller workload: An examination of workload probe.* (DOT/FAA/CT-TN84/24). Atlantic City International Airport, NJ: Federal Aviation Administration.

9 Automation surprises in commercial aviation

An analysis of ASRS reports

Julia Trippe and Robert Mauro

The capabilities of automated flight systems increased rapidly following the introduction of the electronic autopilot in the 1940s. In normal operations, the automated flight system of the modern airliner can now control nearly all functions required for flight. The effect of increased automation has been largely positive, greatly reducing errors due to pilot fatigue and allowing consistent precise navigation and performance. However, automation has given rise to new problems caused by faulty interactions between the pilot and the autoflight system (AFS). This class of problems has been variously termed lack of mode awareness (Javaux & De Keyser, 1998), mode confusion (Degani, Shafto, & Kirlik, 1999), and automation surprise (Wiener & Curry, 1980; Sarter, Woods, & Billings, 1997, Woods & Sarter, 2000; Burki-Cohen, 2010). In these cases, the flight crew expects the automation to command one behavior and is surprised when it commands another. Automation surprise may result from undetected failures in aircraft sensors or other systems. Automation surprise also may result from pilots having an inadequate or mistaken "mental model" of the machine's behavior in the operational environment (Sarter & Woods, 1995). In addition, automation surprise may result from a problematic interface that does not provide adequate information about the status of the machine (Norman, 1990; Feary et al., 1998; Degani et al., 1999).

When they do not jeopardize flight safety, automation surprises are a nuisance. Pilot lore is replete with complaints of flight management systems misbehaving. Flight management computers can appear to be pernicious allies that on occasion unilaterally decide to "drop" fixes, void altitude restrictions, or change modes of operation. These events increase pilot workload, setting the stage for other errors. They create inefficiencies for the aircraft directly involved and may disrupt the flow of air traffic as controllers vector other aircraft to accommodate the aircraft whose crew is dealing with the unexpected behavior. For the most part, the result of these events are relatively benign. No metal is bent; no one is injured; no lives are lost. But this is not always the case. When the automation commands an aircraft trajectory that violates airspace or operational limitations, automation surprise becomes a critical problem (Reveley, Briggs, Evans, Sandifer, & Jones, 2010).

Undetected failures

Unexpected behaviors of the AFS have been implicated in a number of recent fatal accidents. For example, Sherry & Mauro (2014) analyzed nineteen modern airliner Loss of Control (LOC) accidents resulting in aerodynamic stalls. These accidents involved structurally and mechanically sound aircraft decelerating through the $1.3V_{Stall}$ buffer to the stall airspeed (V_{Stall})—that is, a Controlled Flight into Stall (CFIS). All nineteen accidents and incidents can be described by the sequence of events presented in Figure 9.1: A triggering event (e.g., sensor failure) has an effect on the automation (e.g., mode change). This leads to an inappropriate command by the AFS for pitch or thrust. The inappropriate command occurs while the aircraft is conducting an appropriate deceleration maneuver, slowing intentionally to near the minimum safe operating speed, which either remains fixed or increases (e.g., due to ice contamination on the wing). Finally, the flight crew fails to intervene to arrest the deceleration into a CFIS. Although a deceleration through $1.3V_{Stall}$ and then through V_{Stall} occurred in each accident, there is no pattern or consistent failure in the types of triggering events, in the effects of the triggering events on the automation, or in the inappropriate commands generated by the automation. Furthermore, there was no single flight crew response to the events that led to the CFIS accidents that would have been appropriate in all cases.

Failures in the examined CFIS accidents can be considered "functional complexity failures." That is, failures that are due to unanticipated problems resulting from complex interactions between factors in a complex system. In complex systems, the processing and communication within the system are crucial. For example, in several accidents (e.g., Turkish Airlines 1951, Air France 447, XL Germany) a relatively minor malfunction in a sensor triggered the sequence that led to the destruction of the aircraft. In these cases, the automation functioned as designed, either disengaging or continuing to operate seemingly as expected, but using the inaccurate data from the sensors. The concept of operations for the "flight deck system" is for the flight crew to delegate tasks to the automation and to supervise its performance. In the event that the automation generates an inappropriate command (e.g., throttles maintain idle thrust when the crew expects them to advance), the flight crew is expected to diagnose the problem and intervene. However, the ability of the flight crew to detect these rare events in a timely manner and to act appropriately is severely compromised when the required information is not available to the crew, is inaccurate, is not displayed in an easily understood format, or the crew does not understand how the automation operates.

Faulty mental models

To prevent or mitigate the effects of these "automation surprises," one must first understand why they occur. People are surprised when they expect one

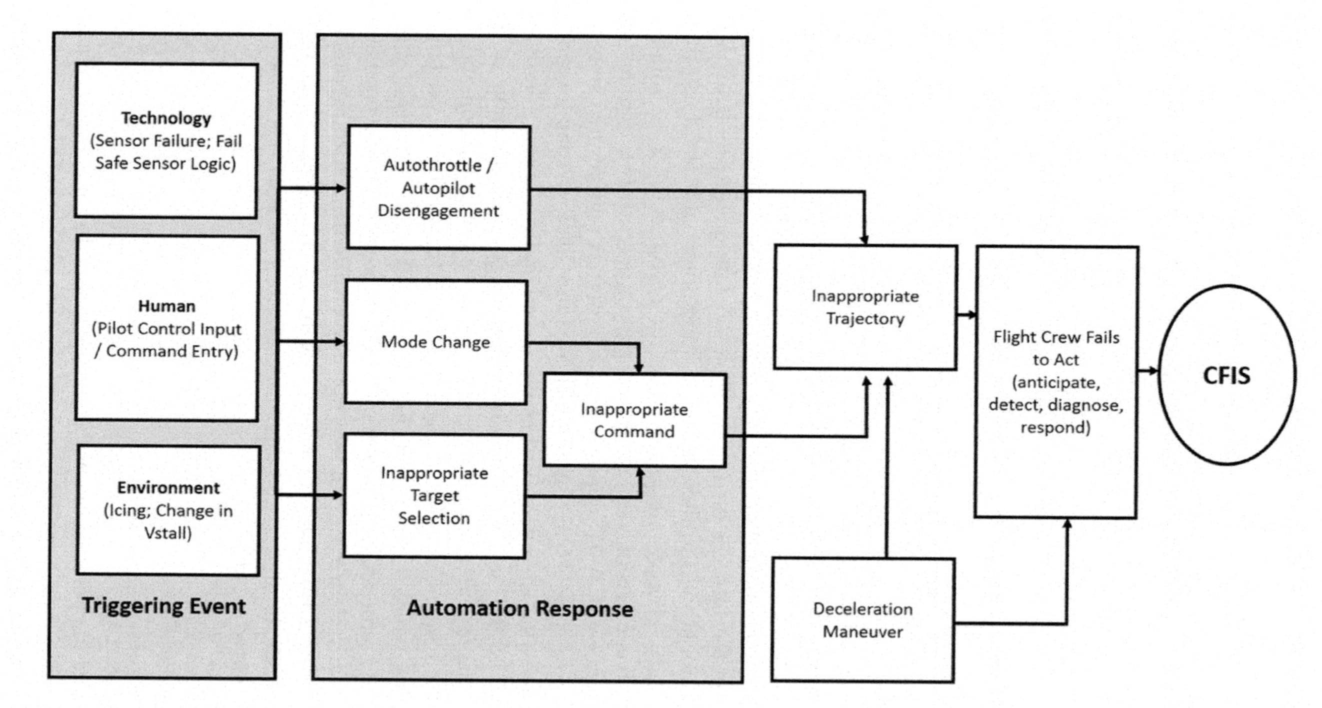

Figure 9.1 Sequence of events leading to CFIS.

event but another occurs. So, to understand automation surprise, one must ask why the behavior of the AFS was not expected. Based on their training and experience, pilots build an understanding (a "mental model") of how their automation functions. Selected information about the current status of the aircraft, including its automation, is interpreted in the context of this mental model and used to build a mental representation of what the aircraft is currently doing and what it will do next. Hence, to be surprised, either the information fed into the mental model must be inaccurate or the model itself must be wrong. Pilots' expectations of what their automation will do may be in error when they attend to the wrong data, misinterpret data, or the data is in error. Alternately, their expectations may be wrong when their understanding of what the automation will do under the encountered conditions is wrong.

In this chapter, we examine a sample of pilot reports of unexpected automation behavior chronicled in the Aviation Safety Reporting System (ASRS) database in an effort to characterize the nature of automation surprise. Because commercial aviation accidents are fortunately very rare, they provide a very limited view of the potential underlying issues. Because a factor has not been implicated in an accident does not mean that it does not present an important safety concern. Voluntary safety reports provide a broader view and offer one of the few sources of data available on infrequent problems in the airspace system. However, these too have limitations and biases. First, the reports are voluntary. Hence, only matters that the reporters think are important are included. Reporters may overlook incidents or potentially important aspects of incidents. Second, ASRS reports provide reporters with limited amnesty from disciplinary actions. Hence, individuals are more likely to submit reports when they believe that the event has been observed and could result in disciplinary action. Third, the sample of published reports is a biased sample of the filed reports. Due to severe budgetary limitations, only a small fraction of the reports that are filed are included in the database. The decision process used by analysts in deciding which reports to include is not clear, but is likely to be heavily influenced by the analysts' judgment of the importance of the problems raised in the report. Hence, one should not rely on frequencies of reports to make direct statistical inferences to rates of events in the airspace system. Rather, these analyses should be used to gain insight into the relations between factors in the reported incidents.

Methods

The ASRS database was searched for automation-related event reports from 2012 by crews operating under Part 121 (scheduled airliners). The initial search criteria were broadly specified to minimize the likelihood that relevant reports would be missed. Reports that mentioned automation, autopilot, auto throttle, flight management system, flight management computer, control display unit, or any of the common abbreviations for these devices were retrieved. Of the 558 reports obtained, 234 described an event in which the pilots were surprised by unexpected actions of the AFS.

The events that transpired before, during, and after the surprise were coded. Distinctions were made between five categories of events (actions or circumstances): precipitating, contributing, problem, detection, and response. *Precipitating* actions were those that preceded and led directly to the automation surprise. Within this category, we distinguished between primary precipitating or "catalytic" actions and secondary precipitating actions that occurred in response to the catalytic events. For example, in a number of cases, Air Traffic Control (ATC) instructions directed pilots to alter their previously programmed flight path. In entering flight path alterations into the flight management system (FMS), the pilots made an error that later resulted in a surprising aircraft behavior. We coded the ATC instructions as the primary catalytic precipitating event and the pilot programming of the FMS as the secondary precipitating action. *Contributing* circumstances were those that did not directly precipitate the automation surprise but that may have contributed to the problem. For example, pilots may have reported being rushed or fatigued during the operation. *Problem* events were those that produced the surprise. For example, in the prototypical ATC precipitated event described above, the pilots were often surprised by the aircraft veering away from the intended course. In this case, the problem event was coded as a course deviation. *Detection* actions were those that led to the discovery of the problem. These actions could involve direct observation of the aircraft behavior (as in the example above) by the pilots or ATC or observation of messages (e.g., Electronic Caution Alert Module (ECAM)), alerts (e.g., autopilot disconnect), or control movements. *Response* actions were those taken to resolve or recover from the surprise. These included actions such as taking manual control of the aircraft, switching to a lower level of automation (e.g., from Vertical Navigation Mode (VNAV) to Mode Control Panel (MCP) control), or notifying ATC of a deviation. For each of these event categories, we coded the nature of the event, when the event occurred (phase of flight), who performed the action (e.g., ATC, crew, AFS), and the level of automation in use.

Results and discussion

What was surprising?

A variety of different automation-related events surprised crews (see Table 9.1). In 15 percent (thirty-five) of the cases, the crew was surprised by changes in AFS operation, including shutdown or freezing of various AFS components. In eleven of these cases, the autopilot disconnected. In three of these cases, the auto throttle disengaged or otherwise behaved unexpectedly.

Table 9.1 Surprising automation-related events

Event type	*Percent*	*Total – N*
AFS operation only	**15.0**	**35**
AFS component failure	4.7	11
Auto throttle	1.3	3
AP disconnect	4.7	11
Unexpected mode change	4.3	10
AFS interface only	**12.0**	**28**
Display	6.0	14
FMS drop	6.0	14
AFS problem affects aircraft control	**9.0**	**21**
AFS problem affects aircraft behavior	**64.0**	**149**
Airspeed	11.5	27
Altitude	14.1	33
Course	15.0	35
Localizer	2.6	6
Vertical path	20.5	48
Other	**0.0**	**1**
Total	**100.0**	**234**

For example, in the following scenario, the pilot of an Airbus 300 flying through light icing reported an autopilot and autothrottle disconnect:

> (ACN: 1055883) Shortly after level off at FL220 in intermittent light icing conditions and light to moderate turbulence, my IAS speed tapes/indication on my PFD went smoothly and expeditiously to around 50 KTS indication. The autopilot kicked off and clicks were heard suggesting other switches and systems were dropping off. The compound situation of turbulence, IMC flight conditions and numerous unknown system losses made for a very interesting 2 minutes of quick evaluation and prudent action on our part as a crew. I initially maintained pilot flying duties. There were no sensory inputs that we had left level, constant speed flight. This (sensory awareness of normal controlled flight path) was lightly brought into question by the frequent change in light to moderate turbulence we were encountering. It amounted to very quickly ensuring what was working and what wasn't by referring to standby instruments and comparing them to primary sources. With my focus on aviating I did not notice that the autothrottles had dropped off (one of the "clicks" we heard I suspect) and the aircraft was slowly accelerating. The combination of my PFD primary speed showing 50–60 KTS, me hand flying in turbulence, the First Officer's IAS tapes on his PFD showing overspeed, my referencing the standby gyro and the alternate airspeed indicator to accurately maintain safe flight trajectory had us loaded for about 2 minutes. The variability of instrument indications presented to us during the short time frame was the challenge. During this high workload 2 minutes and limited confidence

> in what exactly was "correct" and therefore able to rely upon for safe instrument flight I had climbed approximately 350 FT high . . . My speed tape on my PFD remained at 50–60 KTS until descending back through 14,000 on the arrival then everything seemed to return to normal.

In this case, the AFS apparently disengaged and turned control over to the flight crew as it was designed to do upon receiving unreliable airspeed indications. However, in these circumstances, the pilots must then cope with the same problem that the computer was unable to solve and maintain control of the aircraft.

In ten cases, an unexpected mode change occurred and in eleven cases some component of the AFS froze or failed. For example, in the following scenario, the pilot of a Boeing 737-800 on an RNAV (Area Navigation) arrival reported an unexpected AFS mode change (LNAV refers to lateral navigation):

> (ACN: 1024227) We were switched from the OLM Seven to an RNAV arrival right at top of descent, off of a RADAR vector for spacing. We were able to program the last minute change and brief the approach, but time and workload were at a premium at that point, and getting the aircraft back into path was a struggle, we were high and fast. This put us into a full speed brake situation with speed intervention in order to meet the next speed restriction. Next, ATC issued a slow down during this maneuver (250 KTS) which boxed us in even more. Around 10,000, ATC issued a runway change from 34L to 34C. The Captain re-programmed the approach and set me up for 34C, but during this, the FMC caused a reversion to CWS and CWP and I lost any kind of LNAV/VNAV info for a brief period. Automation was soon restored and we continued uneventfully, probably because ATC reverted us to a vector clearance. This practice of issuing speed restrictions and not giving us lower soon enough really messed up our automation and caused too much distraction in my opinion. This could have easily led to a deviation. If we are going to fly RNAV arrivals, we need to be able to fly them as published when they are built to such close tolerance as these new SEA arrivals. We completely negated any kind of fuel savings with all of the speed and altitude changes (lack there of) during the procedure.

In this case, the crew's attempts to reprogram the FMS "on the fly" to comply with changing ATC instructions apparently led to an unexpected mode change and flight plan discontinuity. The user interface for the flight management computer (FMC) is designed to allow the pilot to program and verify the entire flight plan prior to departure. Attempting to alter the programming during dynamic flight frequently leads to input errors and unexpected consequences.

In 12 percent (twenty-eight) of the cases, the crew was surprised by the AFS interface, but the aircraft's behavior was not affected. Half (fourteen) of these cases occurred when FMS data disappeared unexpectedly. However, in fully 73 percent (170) of the analyzed cases, the crew was surprised by aircraft behavior. In twenty-one of these cases, the crew detected unexpected changes in aircraft control prior to a substantial change in aircraft position or velocity. For example, in the following scenario, the Captain of an airliner is surprised when the programmed departure is dropped from the FMC:

> (ACN: 1041395) After loading the route, departure and runway, some subsequent action caused the runway and departure to be removed. We discovered that the runway was missing on taxi and it was reloaded, but we did not see that the departure was missing. LNAV was selected for the departure. We verified the runway and after takeoff a left turn was commanded. That was normal, however, the remainder of the departure was absent and a continued left turn was directed. I quickly determined that I was getting incorrect guidance so reverted to Heading Select. However, in the transition period we turned beyond the prescribed departure course before making the correction.

FMSs differ in how and what they communicate to the pilot. In some cases, the indication that a flight plan is not properly programmed can be quite subtle, amounting to a few characters on the flight mode annunciator (FMA) or a line on a page of the control display unit (CDU) that may not currently be displayed. In this case, the crew failed to notice that the FMC was not following the flight plan that they thought was loaded until the aircraft was off course.

In 64 percent (149) of all cases, the aircraft's velocity or position was altered substantially without the crew noticing. Of these, twenty-seven resulted in airspeed changes, thirty-five in course alterations, thirty-three in altitude deviations, and in forty-eight cases the aircraft's vertical path was affected unexpectedly. For example, in the following scenario, the Captain of an Airbus 300 reported being surprised when the aircraft suddenly pitched over despite careful monitoring of the Standard Terminal Arrival Route (STAR) procedure (in this case an RNAV STAR—a standard arrival route that relies on area navigation).

> (ACN: 1043552) Control: As we approached ARG, we were cleared to descend via the FNCHR1 RNAV STAR. All altitude and airspeed constraints were previously reviewed and confirmed by the crew during the approach briefing. The only modifications to the [line selected] FMS procedure were to change the Cost Index to 132 just prior to top of descent, and the deletion of the "vector leg" after JAYWA in order to provide a more accurate Fuel Prediction figure for landing.
>
> The arrival was flown in Profile with the Altitude Select window set and confirmed for 4,000 FT (The lowest altitude on FNCHR1 arrival at JAYWA). All altitude and airspeed constraints were satisfied for all points from start of arrival through crossing LOONR waypoint. After crossing LOONR at 9,000 and 210 KTS, aircraft continued descent in Profile for

> the "at or above" 8,000 crossing restriction at BOWEN. Passing through approximately 8,700 FT, the aircraft pitched over with rapidly increasing airspeed, though target airspeed still displayed 210 KTS and crossing restriction at BOWEN still showed 8,000 on the FMS and the constraints display.
>
> I initially attempted to control the airspeed increase with speed brakes, but it became apparent that further action would be required to correct the deviation. I then disconnected the autopilot to arrest the descent and increasing airspeed. We crossed BOWEN at approximately 7,700 and 240 KTS. We stopped the descent at 7,500 FT and slowed to 210 KTS before continuing the descent via the FNCHR1 arrival during which Approach cleared us to stop our descent at 6,000.

FMSs often contain hidden modes that are not displayed to the pilots, and pilots are frequently unaware of the priorities programmed into the system by the manufacturer. When will airspeed constraints be sacrificed for altitude limits? How can you tell what the system will do as opposed to what it has been instructed to do?

When were crews surprised?

Overall, 55 percent of the automation surprises occurred during the arrival and approach phases of flight (see Table 9.2). By contrast, only 13 percent of the events occurred during cruise. However, this pattern differed according to the type of event. Failures or freezing of autoflight system components, categorized as "AFS component" failures in Table 9.2, were evenly divided between climb, cruise, and approach. Two of the three autothrottle events occurred in cruise. Although most of the autopilot disconnects, unexpected mode changes, and control anomalies occurred during arrival and approach, a substantial proportion occurred during climb and cruise. Display faults occurred equally during climb, cruise, and arrival. Waypoints dropping from FMS flight plans as well as lateral course deviations occurred mainly during climbs below 10,000 feet, and arrival and approach phases.

There are several possible explanations for the disproportionate number of surprising events reported during the approach and arrival phases of flight. During these transitional phases, crews are preparing for landing, traffic is increasing, and ATC often places additional demands on pilots. To comply with these demands, pilots make heavy use of their automation, resulting in discovery of automation problems that lay dormant during previous phases of flight. Furthermore, pilots may make errors while changing modes and programming flight plans. The ASRS reports provide indirect evidence for these explanations. One common factor between the incidents in initial climb, arrival and approach phases of flight (total of 73 percent of all cases) is that they often occurred while the aircraft was following a performance-based navigation (PBN) RNAV standard instrument departure (SID), or STAR procedure. On the other hand, problems that are likely to be caused by issues with

Table 9.2 Surprising events by phase of flight

Surprising event	*Percentage of events by phase of flight*										
	Before push	*Take off*	*Climb below 10K*	*Climb above 10K*	*Cruise*	*Descent above 10K*	*Arrival*	*Approach*	*Go around*	*Unknown*	*Total – N*
AFS operations											
AFS component	0	9	18	9	27	9	0	27	0	0	**11**
Auto throttle	0	0	0	0	67	0	0	0	33	0	**3**
AP disconnect	0	18	0	0	27	0	36	18	0	0	**11**
Mode change	0	0	30	0	20	0	40	10	0	0	**10**
AFS interface											
Display	0	14	29	0	21	0	29	7	0	0	**14**
FMS drop	7	7	29	7	0	0	7	36	0	7	**14**
Control	0	0	19	10	29	5	14	19	5	0	**21**
Aircraft behavior											
Airspeed	0	0	15	15	15	0	11	37	7	0	**27**
Altitude	0	0	0	12	3	9	36	39	0	0	**33**
Course	0	0	37	6	14	0	20	17	0	6	**35**
LOC	0	0	0	0	0	0	0	100	0	0	**6**
Vertical path	0	0	4	0	2	15	69	10	0	0	**48**
Other	0	0	0	0	0	0	100	0	0	0	**1**
Percentage of total	**0**	**3**	**15**	**6**	**13**	**5**	**31**	**24**	**2**	**1**	**234**

electronic components (e.g., AFS and display problems) were more likely than other problems (e.g., course or altitude deviations) to occur during cruise. This follows, given that electrical failures are likely to be dependent on the amount of time spent in an operation, and airliners typically spend most of the time during a flight in cruise. In contrast, flight path deviations are likely to occur when the aircraft is being commanded to change course or altitude—which is more common during climb, arrival, and approach. This is also when ATC is likely to be directing ad hoc changes in course or altitude that require the crew to depart from their preplanned flight path.

Proximal precursors of surprising events

Narratives of the ASRS reports were analyzed to determine the sequence of events that preceded the automation surprises. Sometimes, the pilots described probable causal sequences. This was particularly likely when the pilots determined that their own actions led to the surprise. In other cases, analysts could reasonably infer an event sequence based on pilots' descriptions of their actions and the automation response in combination with knowledge of AFS operations. Problems attributed to malfunctioning automation components rarely contained sufficient information to verify this conclusion. In many of these cases, pilots reported contacting maintenance, but rarely reported ensuing findings.

A precipitating event could not be determined with sufficient confidence in twenty-one of the examined cases. Of the remaining 213 cases, 66 percent (140) involved human errors in AFS operation (see Table 9.3). Pilot actions led to the majority of airspeed (64 percent), altitude (84 percent), course (71 percent), localizer (60 percent), and vertical path (73 percent) surprises. Pilot actions also led to the majority of surprises resulting from AFS problems, control manipulations, display problems, unexpected mode changes, and dropped waypoints. However, in 24 percent of cases, the AFS or another technological system was apparently responsible for triggering the surprise. These system-induced surprises include two out of three auto throttle changes, the majority (80 percent) of the autopilot disconnect events, and a large proportion of the autoflight problems (44 percent), control manipulations (44 percent), display problems (38 percent), and mode changes (33 percent).

Although 66 percent of the surprising events were precipitated by pilot actions, these actions were unabetted in only a small proportion (28 percent) of these cases. In the remaining 72 percent of these cases, pilot actions were triggered by external events. In 20 percent of the cases, pilots were attempting to cope with equipment issues when they inadvertently triggered an unexpected automation action. In the remaining 52 percent of the cases (seventy-three), pilots were attempting to comply with ATC instructions when they inadvertently triggered the unexpected automation response. Of the reports describing an automation surprise that occurred during an RNAV procedure, 36 percent

Table 9.3 Surprising event by source of event

Surprising event	*Percent of events by source of event*					
	Pilot	*Env.*	*AFS*	*Other sys*	*Other*	*Total – N*
AFS						
AFS component	44	0	33	11	11	**9**
Auto throttle	33	0	67	0	0	**3**
AP disconnect	20	0	20	60	0	**10**
Mode change	56	11	33	0	0	**9**
AFS interface						
Display	54	0	23	15	8	**13**
FMS drop	100	0	0	0	0	**11**
Control	44	11	22	22	0	**18**
Aircraft behavior						
Airspeed	64	28	8	0	0	**25**
Altitude	84	3	6	3	3	**31**
Course	71	0	17	6	6	**35**
LOC	60	40	0	0	0	**5**
Path	73	2	18	2	5	**44**
Percentage of total	66	7	16	8	3	**213**

arose from a sequence in which ATC requested a change of flight path and the crew proceeded to make an error in reprogramming the AFS, resulting in unexpected behavior of the AFS. However, many reports did not stipulate the exact procedure being performed, so we could not determine what percentage of cases occurred during PBN approaches or departures.

Detection

In general, pilots were the first to detect surprising events. In eighty-five cases, the crew noticed the surprising event before it affected aircraft behavior. However, in 149 cases the surprising event was not noticed until the aircraft's behavior was affected. In these cases, either the crew or ATC could have been the first to detect the event. In one of these cases, the first detector could not be determined from the report. In 68 percent of the remaining cases, the pilots were the first to detect the event (see Table 9.4). The crew always detected airspeed deviations before ATC. In 42 percent of altitude deviation cases, ATC detected the deviation simultaneously (18 percent) or before (24 percent) the pilots. In 17 percent of localizer deviation cases, ATC detected the deviation before the pilots. In 17 percent of vertical path deviations, ATC detected the deviation simultaneously (4 percent) or before (13 percent) the pilots. However, in 71 percent of course deviation cases, ATC detected the deviation simultaneously (9 percent) or before (63 percent) the pilots. It is not clear why pilots were relatively less attentive to course than other deviations or alternately why ATC was more attentive to course deviations than pilots. However, a relatively large proportion of course deviations occurred during climb below 10,000 feet. In many of these cases, this reflects lateral deviations from SIDs flown using LNAV when the pilots were unaware that the planned flight plan was not being executed.

Resolution

The automation level at which the aircraft was being operated at the time of the surprising event was compared to the automation level during resolution of

Table 9.4 Who detected surprising event

Type of deviation	*Percentage of events by detector of event*			
	Crew	*ATC*	*ATC & crew*	*Total – N*
Airspeed	100	0	0	**27**
Altitude	58	24	18	**33**
Course	29	63	9	**35**
Localizer	83	17	0	**6**
Vertical path	83	13	4	**47**
Total – N	**100**	**37**	**11**	**148**

Table 9.5 Automation level during surprise event by automation level during resolution

Automation level during surprise event	*Percentage of events by resolution mode*					
	VNAV	*NAV*	*MCP*	*A/P*	*Manual*	*Total – N*
VNAV	32	0	22	4	42	**72**
LNAV	3	56	14	6	22	**36**
MCP	0	0	64	4	32	**28**
A/P	0	0	2	43	54	**46**
Manual	0	0	18	0	82	**17**
Percentage of total	**12**	**10**	**22**	**13**	**44**	**200**

Note. A/P (autopilot) refers to cases in which an unspecified level of automation was in use.

the event (Table 9.5). In thirty-four cases, the automation during one period or the other could not be determined with reasonable certainty. In 48 percent of the remaining 200 cases (ninety-five), the same level of automation was maintained throughout the reported event. In 39 percent of the cases (seventy-two) in which the aircraft was being flown under some level of automation, pilots resorted to manual control following the surprising event. In the remaining 61 percent of these cases (111) automation was used in the recovery. When VNAV was in use at the time of the surprising event (seventy-two cases), pilots resolved the issues and continued under VNAV 32 percent of the time. In 22 percent of the VNAV cases, the crews used MCP inputs to control the aircraft; 42 percent of the crews resorted to manual control. When LNAV was in use at the time of the surprising event (thirty-six cases), the pilots continued to fly using LNAV 56 percent of the time. In 14 percent of the LNAV cases they switched to the MCP. In 22 percent of LNAV cases, pilots resorted to manual control. When the aircraft was being controlled using the MCP at the time of the event (twenty-eight cases), pilots continued to fly using the MCP in 64 percent of the cases and resorted to manual control 32 percent of the time.

General discussion

Three important conclusions can be drawn from the results discussed above. First, many different factors may precipitate automation surprises. These include problems in the AFS, problems in the displays and interface with the automation, problems in other aircraft sensors and systems, interactions with weather and other aspects of the external environment, and inappropriate actions taken by the pilots. Hence, no single technical solution is likely to succeed in substantially reducing the number of automation surprises. Second, inappropriate actions by pilots are involved in a large proportion of the automation surprise events. However, in the majority of these cases, these inappropriate actions were made by pilots attempting to cope with ATC instructions that required unplanned changes to their flight plans. Reducing the number of in-flight

flight plan changes is likely to reduce the number of automation surprises. Third, in most cases, pilots continued to fly successfully using some level of automation following the automation surprise. Automation can relieve pilot workload and allow precise handling of the aircraft not typically attainable with manual control. Recovery from automation surprises need not require reversion to manual control.

Although there is no single general intervention that can prevent automation surprise or completely mitigate its effects, several different tactics used together may prove effective.

Training methods

A large proportion of the reported automation surprises can be traced to inappropriate pilot actions. In some cases, the pilots came to understand what had caused the surprise after the fact. In other cases they did not, but the probable causes of the surprises were apparent in their reports. The number of inappropriate actions could be reduced if pilots possessed a more thorough understanding of their aircraft's automation. Manufacturers of automated systems have long touted the ability of automation to simplify pilots' tasks and improve precision and efficiency. However, researchers have repeatedly noted that while aviation automation has improved the efficiency and precision of operations, it has not reduced complexity. Indeed, automation may have increased the complexity of the pilot's job. Pilots often plead for manufacturers to make the automation simpler, and there may be modifications to interfaces that would help. However, the complexity of the automation itself cannot be substantially reduced. It must be complex because the operational environment is complex and dynamic. But training has not kept pace.

One response to training complex automation has been to reduce complexity by limiting the training to the mechanics of executing particular procedures and limiting pilot discretion to the execution of only these procedures. However, this strategy limits pilots' understanding of the automation. When conditions arise that do not correspond to those covered by the trained procedures, the actions of the automation may surprise the pilot who has no deeper concept of how the automation operates. Methods for automation education need to be developed which can help pilots develop an understanding of their automation that allows them to anticipate automation actions and not simply respond with a small set of canned procedures. For pilots to construct adequate mental models of automation, they do not need to know the intricacies of the underlying engineering, but they must know how the system interacts with the environment—how it obtains information, what it controls, and what targets it is trying to achieve. Hence, for each automation mode, pilots must be trained to understand: (1) what is being controlled (e.g., pitch, thrust); (2) what data about the current state of the aircraft is being used (e.g., altitude from the Captain's radio altimeter, lateral position from GPS (Global Positioning System)); (3) what targets are being pursued (e.g., altitude from the MCP, speed from the FMC); and (4) what actions will be taken when the targets are achieved (e.g., proceed on heading, revert to programmed flight plan).

But having this knowledge is not sufficient. It merely provides the framework for the mental model. At every point during a flight, the model must be populated with current information about the state of the aircraft and how it relates to the intended flight path. This requires that pilots be trained to: (1) know where to find the relevant information; (2) attend to these sources; (3) interpret the information correctly; and (4) integrate this information with their stored knowledge of the AFS and intended flight path.

The AFSs in current use frequently do not present all of the relevant data in a single easily interpretable display. The requisite data may be scattered between MCP windows, the FMA, the navigation display, and one or more pages of the CDU. Furthermore, understanding what these data mean for the operation of the aircraft requires retrieving from memory knowledge that may not have been thoroughly learned in the first place.

Flight deck displays

Improved displays should be developed that provide pilots with predictive indications. As in the CFIS accidents described earlier (Sherry & Mauro, 2014), in most of the ASRS cases examined here, the automation performed as it had been programmed to perform. However, there was a discrepancy between what the pilots thought the system was programmed to do and what it was actually programmed to do. Typical automation interfaces do not provide clear displays of the programmed and predicted speeds, or vertical and lateral flight paths.

One partial solution is the intentional FMA (Feary et al., 1998; Sherry & Polson, 1999). Because of its limited utility, ambiguous symbology, and the fact that in many cases automation intentions can be accurately inferred from other indications, pilots tend to spend relatively little time attending to the traditional FMA (Mumaw, Sarter, & Wickens, 2001). However, as is apparent from the CFIS incidents, there are situations in which the same actions can be the product of different automation intentions. Instead of relying on pilot inferences, automation intentions could be displayed on a modified FMA that clearly displays both the current state of the automation and its intentions. Experimental results indicate that significant improvements in flight crew awareness of automation actions can be achieved through the annunciation of automation intentions (Feary et al., 1998).

ATC procedures

A large portion of pilot precipitated automation surprise events were themselves caused by instructions from ATC that proved problematic for pilots. A substantial decrease in the number of automation surprise events likely could be attained through restructuring ATC arrival and approach policies to decrease the number of unnecessary "mission surprises" with which pilots must cope. Because these complex approaches effectively must be flown by automation, modifications force pilots to program the FMS while flying the procedure. In this process, errors may be made that surprise the pilots and disrupt the flow of traffic. Modifying ATC methodology in handling these procedures could substantially decrease the number of these problems. In a large number of the cases examined in this study, ATC was the first to detect a flight path deviation. Frequently, ATC handled the problem by providing a new clearance. In the current airspace system, these deviations present a safety hazard. Under NextGen, aircraft may fly in closer proximity along defined 4D paths. Under NextGen, deviations such as those observed in the current study would bring aircraft dangerously close to one another. At best, these events would cause substantial disruption to the traffic flow. At worst, they could result in collisions.

Many of the analyzed events occurred while on an RNAV STAR or SID. Some RNAV arrivals are designed with complex vertical and lateral paths tailored to the terrain and traffic requirements of individual airports. These procedures enable aircraft to fly efficient lateral paths and use reduced or idle power settings for their entire descent, rather than leveling out intermittently (Joint Planning and Development Office, 2007). However, these procedures may be so complex that they cannot be reliably flown manually. For example, the FRDMM THREE RNAV STAR with BUCKO transition into the Washington, DC area (FRDMM and BUCKO are two waypoint names in the procedure) has 15 GPS defined waypoints covering 140 miles (see Figure 9.2 for the first portion of this procedure). At typical approach speeds, these waypoints lie 1.2 to 3.6 minutes apart. Each fix has an associated altitude restriction that can take the form of a window, a ceiling, a floor, or an exact altitude. Each fix may also have an associated heading change and/or an airspeed restriction.

Flying complex RNAV procedures without "real" VNAV (computer controlled vertical and lateral navigation with control over aerodynamic controls and autothrottles) is extremely difficult. If flight path changes are required during a VNAV segment, reprogramming while continuing on course can cause a multitude of automation problems, as seen in the ASRS reports cited above. Attempting to fly PBN procedures with less sophisticated equipment creates additional problems reflected in higher workload for flight crews and increased potential for problems. However, carriers operating aircraft without this level of AFS sophistication (e.g., without autothrottles) may go to great lengths to try to gain authorization to fly PBN procedures. As a result, there is a wide range of equipment as well as capability and comfort in crews' performance of these procedures. This underscores the importance of understanding and developing strategies for addressing the problem of automation surprise before NextGen becomes fully operational.

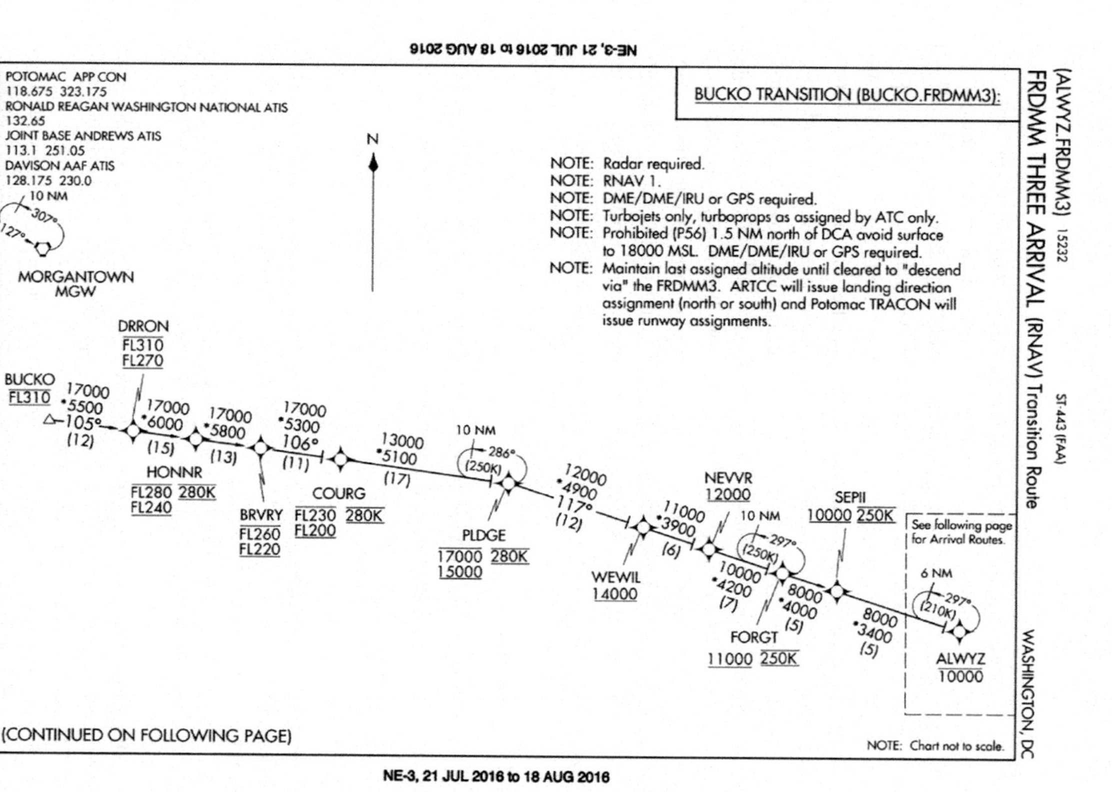

Figure 9.2 FRDMM three arrival RNAV instrument procedure.

Conclusion

We are surprised when our expectations do not match reality. Fundamentally, there are two ways to avoid being surprised: change our expectations to meet reality or change reality to match our expectations. Automation surprises us when it interacts with the external environment in ways that we do not expect. To reduce automation surprises, we can seek to alter the automation, the external environment, or our expectations of how the automation operates. In principle, all of these approaches could be effective and each of these is reflected in the suggestions presented above—training to deepen pilot understanding of the automation, changing the way that the automation communicates with the crew, and altering the operational environment to limit the circumstances that lead to automation surprises. These interventions and others like them may prove effective in reducing the number of automation surprises of the sort described in this chapter. However, we are only at the beginning of a revolution in artificial intelligence. The quasi-intelligent systems that are in operation today will be replaced by systems capable of substantially more complex "mental" activities and autonomous or quasi-autonomous operation. We must develop effective interfaces and methods for learning how to understand, live, and work with these systems, or risk being unpleasantly surprised.

Acknowledgements

This work was funded by NASA NRA NNX12AP14A. Special thanks to Lance Sherry, Immanuel Barshi, and Michael Feary for technical suggestions.

References

Burki-Cohen, J. (2010). Technical challenges of upset recovery training: Simulating the element of surprise. *Proceedings of AIAA Guidance, Navigation, & Control Conference*: Toronto, Ontario, Canada, http://dx.doi.org/10.2514/6.2010-8008.

Degani, A., Shafto, M., & Kirlik, A. (1999). Modes in human-machine systems: Constructs, representation, and classification. *International Journal of Aviation Psychology*, *9*, 125–138.

Feary, M., McCrobie, D., Alkin, M., Sherry, L., Polson, P., Palmer, E., & McQuinn, N. (1998). Aiding vertical guidance understanding. *NASA Technical Memorandum NASA/TM-1998-112217*, Ames Research Center, Moffett Field, CA.

Javaux, D., & De Keyser, V. (1998). The cognitive complexity of pilot-mode interaction: A possible explanation of Sarter and Woods' classical result. In *Proceeding of the International Conference on Human-Computer Interaction in Aeronautics* (pp. 49–54). Montreal, Quebec: Ecole Polytechnique de Montreal.

Joint Planning and Development Office. (2007). *Concept of operations for the next generation air transportation system*. Washington, DC: Government Printing Office.

Mumaw, R., Sarter, N., & Wickens, C. (2001). Analysis of pilots' monitoring and performance on an automated flight deck. *Proceedings of the 11th International Symposium on Aviation Psychology*, Columbus, OH, March 5–8, 2001.

Norman, D. A. (1990). The "problem" with automation: Inappropriate feedback and interaction, not "over-automation." *Philosophical Transactions of the Royal Society B: Biological Sciences*, *327*(1241), 585–593.

Reveley, M., Briggs, J., Evans, J., Sandifer, C., & Jones, S. (2010). Causal factors and adverse conditions of aviation accidents and incidents related to integrated resilient aircraft control. *NASA Technical Memorandum* NASA/TM-2010-216261.

Sarter, N. B., & Woods, D. D. (1995). How in the world did I ever get into that mode? Mode error and awareness in supervisory control. *Human Factors: The Journal of the Human Factors and Ergonomics Society*, *37*(1), 5–19.

Sarter, N. B., Woods, D. D., & Billings, C. E. (1997). Automation surprises. *Handbook of Human Factors and Ergonomics*, *2*, 1926–1943.

Sherry, L., & Mauro, R. (2014). Controlled flight into stall (CFIS): Functional complexity failures and automation surprises. *Proceedings of the Integrated Communications Navigation and Surveillance Conference*, D1-1. IEEE.

Sherry, L., & P. Polson (1999) Shared models of flight management systems vertical guidance. *The International Journal of Aviation Psychology 9*(2), 139–153.

Wiener, E. L. & Curry, R. E. (1980, June). *Flight-deck automation: Promises and problems* (NASA-TM-81206). Moffet Field, CA: NASA.

Woods, D., & Sarter, N. (2000). Learning from automation surprises and "going sour" accidents. In Sarter, N., & Amalberti, R. (eds.), *Cognitive Engineering in the Aviation Domain*. LEA: Mahwah, NJ.

Part IV

Eye and touch in the aviation environment

10 Developing and validating practical eye metrics for the sense-assess-augment framework

Matthew Middendorf, Christina Gruenwald, Michael A. Vidulich, and Scott Galster

So far, the field of human factors has generally focused on creating better controls and displays to help the human operator understand the situation and perform the required actions. In contrast to the impressive system performance gains that have been accomplished by such work (e.g. Wickens, Hollands, Banbury, & Parasuraman, 2013), there has been relatively little work on how to help the system respond better to the human's needs. The present chapter examines eye measurements that could potentially inform the machine side of the human-machine system about the level of mental workload experienced by the human operator, boosting the machine's ability to aid the human adaptively. To realize this potential, the present work describes algorithms that were developed to detect eye blinks and saccades during real-time mission performance.

During the real-time performance of a mission, the traditional approach requires the human to adapt to the machine side of the system. Originally, the machine was merely a conduit for executing whatever the operator commanded; the machine provided little to aid the human's understanding or selection of goals other than providing warnings of potential dangers, such as the "check engine" or "low fuel" lights in automobiles or the "stall warning" indicators in aircraft. However, the role of the machine in the human-machine team has been changing as the automation embedded in the machine has gained more and more intelligence by virtue of increased computational power and sophisticated software. Certainly, more sensors are now available for the machine to better understand the current situation and provide a larger repertoire of danger warnings to the human. For example, in modern automobiles "blind spot" and "lane change" warnings are now available for helping the human driver to operate more safely.

Although this improved situational awareness of more intelligent machines could certainly improve overall system performance and safety, further improvements seem likely to result from providing the machine with better information regarding the cognitive state of the human operator. This is not a new insight. For example, in 1981 the McDonnell Douglas Astronautics Company, under the sponsorship of the Cybernetics Technology Office of the

Defense Advanced Research Projects Agency (DARPA), hosted the ambitious "Biocybernetic Applications for Military Systems" conference (Gomer, 1981). There, biocybernetics was defined as "a real-time communication link from the operator to the system he controls, which uses physiological signals that are recorded as the operator performs assigned tasks" (p. 2). The work presented at the conference tended to highlight the potential role of electrophysiological brain assessments, including event-related evoked potentials. But a number of other physiological data sources such as eye movements and pupillary responses were also considered. Although the work reviewed at the Biocybernetic conference did not lead to any immediate real-world applications, it was representative of a vision to make machines better teammates for their human operators.

One of the most ambitious attempts to advance this vision was the "Augmented Cognition" program headed by Dylan Schmorrow (e.g., Schmorrow & Reeves, 2007; Stanney, Winslow, Hale, & Schmorrow, 2015). Among the agencies that participated in this program were the National Science Foundation, the National Institutes of Health, the DARPA, the Office of Naval Research, and the Air Force Research Laboratory (AFRL). The state of the art technologies required to enable augmented cognition were discussed and critiqued at many meetings, and viable paths forward were debated. These meetings succeeded in building a critical mass of interest and shared knowledge regarding the viability of building such advanced systems.

To capitalize on this momentum, AFRL initiated the Sense-Assess-Augment (SAA) program (Galster & Johnson, 2013; Parasuraman & Galster, 2013) to combine relevant technologies with the goal of producing robust human-machine systems that will be ready to make the transition from laboratory to the operational community.

The SAA framework

The SAA framework is designed to *sense* a suite of physiological signals from the operator, use these signals to *assess* the operator's cognitive state, and *augment* performance to optimize mission effectiveness. The AFRL SAA program aims to combine the necessary components of an intelligent and adaptive human-machine teamwork system within one cohesive research program. The strategy is to leverage available technologies as much as possible and identify the gaps requiring solutions in order to apply those technologies effectively. For example, in regard to the *sense* portion of the SAA framework, there are many off-the-shelf sensing technologies that can provide useful real-time, or near real-time, data streams regarding the human operator's psychophysiological state. These include heart measures, skin conductance, electroencephalogram (EEG), and eye-based measures, among others.

The physiological signals acquired from the human are of little use in their raw state. They need to be processed by specialized algorithms to extract features that have known psychological relevance for cognitive state assessment. For example, heart rate variability can be extracted from the electrocardiogram

(ECG) and eye blink rate can be extracted from the electrooculogram (EOG). These features, and others, can be processed to gauge mental workload in the *assess* portion of the SAA framework. The information then could be provided to the operator as bio-feedback, or displayed to a commander who could take action when a suboptimal level of mental overload is indicated. The information also could be provided to the machine for invoking the appropriate form of *augmentation* to help manage mental workload and improve system performance.

The goal of the current research was to develop robust algorithms for processing signals associated with two types of eye activity for potential use in the SAA framework; specifically, algorithms to detect eye blinks and saccades are explored.

Ocular metrics of eye blinks and saccades

Because vision is paramount for acquiring information in many complex, real-world tasks, the potential of a number of eye movements and blink characteristics to inform the mental workload incurred has been explored for quite some time (e.g., Stern, Walrath, & Goldstein, 1984; Kramer, 1991). Both spontaneous eye blinks (also known as endogenous eye blinks) and saccades have been demonstrated to be sensitive to manipulations of mental workload in the laboratory (e.g., Halverson, Estepp, Christensen, & Monnin, 2012; Chen & Epps, 2013; Di Stasi, Marchitto, Antolí, & Cañas, 2013; Gao, Wang, Song, Li, & Dong, 2013) and in studies utilizing high-fidelity simulations of real-world aviation tasks such as air traffic control (Ahlstrom & Friedman-Berg, 2006).

Spontaneous eye blinks

The spontaneous eye blink is characterized by the absence of an identifiable eliciting stimulus (Stern et al., 1984) and is distinguished from other blinks that are reflexive or voluntary. Reflexive blinks often occur as a protective response (e.g., form an object moving toward the face). Voluntary blinks are the willful brief closure of the eyes. The spontaneous eye blink has a characteristic rate, waveform, and temporal distribution that have been found to be associated with cognitive state variables. These characteristics also discriminate them from grimaces or other facial activity and other lid activity (such as non-blink eye closures).

Blink rates tend to decrease with more demanding tasks regardless of input modality, though visual input tends to result in the greatest suppression of blinks. Increase in visual demand is also associated with blinks of shorter closure duration. Spontaneous eye blinks are coordinated with oculomotor activity in such a way as to minimize interference with information intake. Consequently, blinks tend to occur during moments of low attention demands, minimizing the temporary cessation of information intake. For example, blink rate was found to decrease when the co-pilot took command of an aircraft

(Stern & Skelly, 1984) and when subjects transitioned from a single to a dual task (Sirevaag, Kramer, De Jong, & Mecklinger, 1988).

However, blink rate also has been observed to increase when the navigational demands of a simulated flight mission were increased (Wierwille, Rahimi, & Casali, 1985). One possible explanation to account for the inconsistencies observed might be the types of demand manipulated in the different studies. For example, rapid fluctuation in attention in fast-paced discrete-trial procedures may result in an overall increase in blink rate. That the blink behavior could differ according to the nature of the task demand implicates a level of voluntary control that is not just reflexive.

Among the eye measures that have been found to be reliably informative about the cognitive state of the human operator (e.g., Fitts, Jones, & Milton, 1950; Yang, Kennedy, Sullivan, & Fricker, 2013), EOG data are particularly suitable for use in operational contexts. The sensors required to collect the EOG data are relatively simple. The electrical signal of a blink can be acquired using as few as a single pair of electrodes positioned above and below one eye. Blinks are indicated by the potential difference between the cornea and the retina as the eyelid moves between open and closed positions. Because EOG can be recorded through the use of portable amplifiers, advanced video techniques could be used to permit the ambulatory operator full range of motion during recording. In contrast, other eye measures such as pupil size and eye-tracking data that are highly sensitive to the operator's movement are considerably more challenging to transition to real-world applications. For example, pupil size and eye-tracking data are often acquired using off-body cameras, which can impose a restricted field-of-view. A camera-based system can also lose its lock on the eye if the operator changes seating position or turns his or her head. In contrast, the EOG data needed for blink and saccade detection remain continuous regardless of operator movement.

Saccades

Saccades are fast ballistic movements of the eye to allow the human to focus on a new point of interest. A saccade is followed by a fixation—a period of time when the eye is relatively stationary and useful visual information is gathered. Because visual acuity decreases rapidly away from the current direction of gaze, saccades are required to point the eye at regions of interest. In a scene that contains multiple possible targets, selecting the target for the next saccade involves an interplay between the visual properties of the environment and the goal of the observer. That is, saccades may not be target elicited, but rather are internally generated. They also may be exploratory and not always directed to a specific target.

In normal viewing, several saccades are made each second. There is a consistent relationship between the duration, peak velocity, and amplitude (size) of saccadic eye movements, known as the main sequence. Saccadic duration increases as the amplitude (size) of the saccade increases, and the peak velocity increases with the amplitude. Most saccades have an amplitude of less than

15 degrees, after which the head becomes involved in the movement to redirect the eye. It is important to note that even during a fixation, the eyes are not completely still. Three types of miniature eye movements (tremor, drift, and micro-saccades) occur during fixation to reduce neural adaptation and prevent fading of the visual image (Tokuda, Palmer, Merkle, & Chaparro 2009).

Similar to blinks, saccades have been demonstrated to be useful in cognitive workload assessment (e.g., May, Kennedy, Williams, Dunlap, & Brannan, 1986; Di Stasi et al., 2013). For example, Benedetto, Pedrotti, and Bridgeman (2011) found that as more time-shared tasks were added in a driving environment, the number of exploratory saccades increased. Similarly, Bodala, Ke, Mir, Thakor, and Al-Nashash (2014) and Di Stasi et al. (2010) showed that saccadic velocity increased when workload increased. Also, Marshall (2007) found that saccades provided additional discrimination among levels of cognitive engagement in problem solving, single- versus multi-task driving, and fatigued versus alert visual search beyond that available from blink and pupil size data.

Direct and indirect use of the blink and saccade measures

Blink and saccade measures can be used in two different ways: directly and indirectly. A direct use is to treat them as "signals" that signify a certain cognitive state such as the intensity of mental workload. In contrast, an indirect use is to consider them as "noise" that can contaminate other physiological signals such as EEG. The algorithms used to detect blinks and saccades are proposed to also be able to serve the indirect purpose.

An example of indirect use is the removal of the EEG portions suspected to be contaminated by blinks and saccades prior to interpreting the EEG data for mental workload assessment. The EEG is a noninvasive electrical sensing technique that uses electrodes placed on the scalp to measure brain activity. Depending on the research goals, different sites on the scalp may be used. EEG has been demonstrated to be sensitive to changes in mental workload (e.g., Hankins & Wilson, 1998), and the possibility of using it in cognitive assessment or to guide augmentation is very attractive (e.g., Gomer, 1981; Schmorrow & Reeves, 2007; Stanney et al., 2015). However, EEG signals are easily corrupted by a number of artifacts. For example, in addition to the brain's electrical activity recorded at the scalp, EEG signals can include contaminating potentials from saccades and blinks (Gevins & Smith, 2003). A great deal of research has been directed toward artifact mediation (e.g., Fatourechi, Bashashati, Ward, & Birch, 2007). Common methods of dealing with artifacts in the EEG are artifact avoidance, artifact rejection, and artifact removal.

The blink and saccade detection algorithms use a new technique for artifact mediation referred to as artifact separation (Credlebaugh, Middendorf, Hoepf, & Galster, 2015). With this technique, EEG data analyzed in the time domain are transformed into the spectral domain with flags that can indicate the presence of blinks and/or saccades. A consumer of the EEG spectral results could then use these flags to decide what to do with potentially contaminated data.

General approach to designing blink and saccade detection algorithms

Accurate detection of blinks and saccades is essential regardless of whether they are being sought as direct measures of cognitive states, or as indirect indicators to be used for improving the quality of other data. Robust algorithms for the real-time detection of blinks and saccades are needed, especially for operational applications. The general approach here is to extract defining features of blinks and saccades in the EOG data. To provide computational efficiency and flexibility, blink and saccade detection algorithms were implemented using state machines (Gill, 1962). A state machine is a model of computation that at any given time can be in only one of a finite number of states. A change from one state to another is initiated by a triggering event or condition. The state machine monitors the raw EOG signal to sequentially capture characteristics of the signal, such as slope up at the midpoint, amplitude, and slope down at the midpoint for a blink. The aim of the current work is to develop detection algorithms for eye blinks and saccades that have a high degree of accuracy, are dynamically adaptive, work in real time, and can be calibrated to take into account intra- and inter-individual variations.

Eye blink detection

Basic waveform of a blink

Figure 10.1 presents the basic waveform of a blink signal in the VEOG. Andreassi (2007) describes the waveform as a sharp rise in amplitude immediately followed by a sharp fall. The duration is short, the peak is rounded, and there is a noticeable overshoot before the signal returns to zero. In the discussion that follows, each time the VEOG signal goes above and returns below a threshold value is referred to as a "bump." These bumps include both blinks and non-blinks (i.e., eye movements and noise). Despite the visually distinct waveform of a blink, creating an algorithm to automatically detect these blinks has been challenging due to the varied parameters and noisy signals. Although

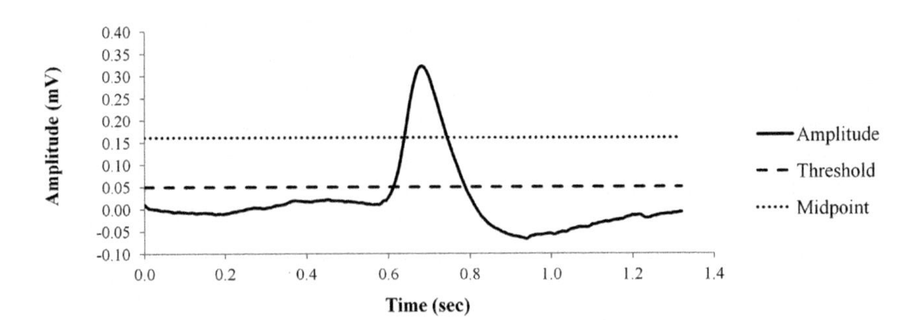

Figure 10.1 The basic shape of a blink.

there exist a number of blink detection algorithms that have been developed for a variety of purposes (e.g., Kong & Wilson, 1998), the detection algorithms that have been reported in the literature lack sufficient detail to guide implementation in real-time applications.

Eye blink detection algorithm components

The blink detection algorithm described here was designed for detection of spontaneous eye blinks, although the other types of eye blinks (reflexive or voluntary) may also be detected. The algorithm also was designed to minimize the contaminating effects of eye movements and other noise. The algorithm has four major components: threshold determination, feature extraction using a state machine, and classification of blinks versus non-blinks.

Threshold determination

The threshold determination algorithm focuses on a sliding five-second window of raw VEOG data. To minimize the effects of blinks and other eye movements on the threshold, the data are high pass filtered using a first order Butterworth filter with a break frequency of 10 Hz. The filtered signal is then rectified and the median is taken for the *raw* threshold value. The median is used because the data in the five-second window can be highly skewed when there is a blink in the window. A second stage of threshold determination imposes limits on the *raw* threshold and incorporates a threshold reduction value to accommodate double and multiple blinks.

Initially, the threshold limits remain constant at a predetermined representative value. After ten blinks have been detected, the limits are dynamic and can be adjusted based on the mean amplitude of previously detected blinks. The threshold reduction value is necessary due to the high pass filter in the signal acquisition hardware (Clevemed BioRadio 150), which causes the signal to overshoot zero on the down slope of the blink. The threshold reduction value is based on the amount of overshoot of the previous blink. The threshold returns to its normal (non-reduced) value using a function that is the inverse of the high pass filter implemented in the signal acquisition hardware.

Feature extraction state machine

The feature extraction state machine uses the threshold to monitor the VEOG signal. The state machine has four states (0, 1, 2, and 3). In state 0 the logic waits for the signal to be below threshold. In state 1, it waits for the signal to go above threshold, at which time upward threshold crossing data are captured and the threshold is frozen. In state 2, the logic waits for the signal to go back below threshold. During this time, peak amplitude and downward threshold crossing data are captured and the threshold is unfrozen. Also in state 2, all data points above threshold (the crown of the bump) are stored to facilitate midpoint calculations. In state 3, the signal overshoot value is captured and the extracted features are scored to see if the signal excursion above and below threshold is a blink. The state machine then returns to state 0.

Blink classification

Software was written to extract a number of features of the bumps that are characteristic of an eye blink for blink classification. The blink detection algorithm is designed to detect four primary features of a blink, as identified by Andreassi (2007). They included the peak amplitude, slope up at midpoint, slope down at midpoint, and duration between midpoints of slope up and slope down (duration for which the bump is above midpoint, see Figure 10.1). Subsequently, to refine the accuracy of the blink detection algorithm, five additional secondary features were also extracted for blink classification. The procedures for establishing criterion values for each of these features and classifying blinks are now described.

Primary criteria. A database of VEOG data from twelve participants performing vigilance and memory (n-back) tasks in an unpublished study was used to establish initial criterion values. Of the 3,102 visible bumps in this dataset, human raters who were trained to recognize the shape of a blink (Figure 10.1) coded 2,020 bumps as blinks and 1,082 as non-blinks. It is common practice to use human raters in blink detection algorithm development (e.g., Divjak & Bischof, 2008; Sukno, Pavani, Butakoff, & Frangi, 2009; Arai & Mardiyanto, 2010; Chen & Epps, 2013; Jiang, Tien, Huang, Zheng, & Atkins, 2013; Toivanen, Pettersson, & Lukander, 2015). The 98th percentile of the minimum and maximum value of each feature were then established. Based on testing, a few adjustments were made to improve accuracy and the criterion values are presented in Table 10.1.

Initially, only bumps that had values which fell within range for all four features would be considered a blink. This initial algorithm was tested on a different unpublished set of VEOG data collected while participants played a variety of computer games. The results were promising, but there was a tendency toward false positives. Some spurious bumps as classified by the human raters were classified as blinks by the algorithm. This indicated a need for further algorithm refinement.

Secondary criteria. To improve classification accuracy, it was decided to extract five secondary features from each bump. The five secondary features were the closure duration, the R^2 values for two (slope up and slope down)

Table 10.1 Criteria values for the primary features

Blink feature	*Minimum criteria value*	*Maximum criteria value*
Amplitude (mV)	0.211	0.6483
Blink duration at the midpoint (s)	0.06	0.198
Slope up at the midpoint (mV/s)	1.5	13.41
Slope down at the midpoint (mV/s)	−10.0	−1.25

Table 10.2 Criteria values for the secondary features

Blink feature	*Minimum criteria value*	*Maximum criteria value*
Closure duration	0.01	0.10
Slope up at the midpoint R^2	0.996	N/A
Slope down at the midpoint R^2	0.995	N/A
Blink duration ZCMP	0.1162	0.3
Blink duration ZCT	0.1	0.35

Note. ZCMP is duration at the zero crossing due to midpoint extrapolation. ZCT is duration at the zero crossing due to threshold extrapolation

linear fits at the midpoint, blink duration at the zero crossing due to midpoint extrapolation (ZCMP), and blink duration at the zero crossing due to threshold extrapolation (ZCT). The two linear fits at the midpoint are extrapolated and the distance between the two fits at the peak is referred to as the closure duration. The distance between these two extrapolations at the zero crossing are used to compute ZCMP. A similar duration (ZCT) is measured using extrapolation of linear fits about the threshold.

The criteria values for these secondary features were determined in the same manner as the primary criteria. The resulting values are listed in Table 10.2.

Final blink classification rules

The classification of any bump as an eye blink is based on the combined assessment of the primary and secondary criteria. One point is awarded when each of the four primary criteria are met and one tenth of a point is awarded when each of the five secondary criteria are met. Therefore, the maximum score for a VEOG bump is 4.5 points. Bumps that score 4.3 points or higher are classified as blinks. This requires that all four of the main features be met, and at least three of the five secondary features be met.

In addition to the classification rule above, additional checks are used to minimize misses and false positives. Bumps that fail only one of the four main criteria, but otherwise have a nearly perfect score (3.4 and 3.5) are given a second look. For example, when a bump fails the maximum amplitude criterion, the criterion could be adjusted upward using amplitude data from previous blinks (minimum of ten required). Similar adaptive tests are applied when the minimum amplitude or the slope down at the midpoint fails. In addition, bumps that have two peaks are rejected.

Validation test

The blink detection algorithm was validated with data obtained from a recent study (Hoepf, Middendorf, & Galster, in preparation). In this study, task demand was manipulated while participants performed a tracking task implemented in a remotely piloted aircraft (RPA) simulation. Participants were instructed to track one or two high value targets (HVTs) by continuously clicking on the video feeds while the HVTs traveled by motorcycle. Experimentally manipulated were the number of HVTs to track (one or two), route taken by HVT (country or city), and weather (clear or hazy conditions). Selection of these manipulations was based on interviews with real-world RPA operators. The operators indicated that tracking two targets would be harder than tracking one, tracking targets in the city would be harder than in the country, and tracking in hazy conditions would be more difficult than in clear conditions.

The VEOG data were processed using the blink detection algorithm. Reduced blink rate and shortened blink duration were observed with increased workload. To directly assess the accuracy of the blink detection algorithm, eight participants were video recorded using a Basler high speed camera while performing two trials in the study. Two human raters independently watched the video recordings to identify blinks. Any disagreements between the two raters regarding classification were resolved by reexamination and discussion. Compared to the human-classified blinks, the blink detection algorithm had 2.3 percent misses and 1.0 percent false positives. Overall, the blink detection algorithm had an accuracy rating of 96.7 percent.

Summary

The blink detection algorithm has several desirable attributes that would enable transition to real-time, real-world applications. This blink detection algorithm does not require baseline data or calibration. There is no need (or mechanism) for experimenter adjustments. Furthermore, it could work for all individuals without calibration. Also, after the algorithm has compiled statistics on a few blinks, it can adapt its criteria to improve classification accuracy. In addition to the algorithm being adaptive, it is dynamic in the sense that the detection threshold will change in real time in response to changes in the VEOG signal. That the blink detection algorithm has been validated to produce measures that were sensitive to workload manipulations supported the algorithm's viability as an assessment component of the SAA paradigm to guide potential augmentations in future research (Hoepf, Middendorf, Epling, & Galster, 2015).

Saccade detection

The saccade detection algorithm presented here is a novel approach because it uses polar coordinates. This means that saccades are reported in magnitude and angle. To detect saccades using polar coordinates requires four electrodes; two for the horizontal electrooculogram (HEOG), and two for the vertical electrooculogram (VEOG).

Saccade detection algorithm components

The seven major components of the saccade detection algorithm are signal filtering, threshold determination, saccade endpoint detection, saccade startpoint detection, mathematical calculations, classification, and distinguishing saccades from blinks.

Signal filtering. The signal filtering of the raw EOG data stream is necessary to improve the accuracy of the measured saccade amplitude. The raw EOG contains saccades that are evident to the naked eye (Figure 10.2). The distinctive shape of a saccade contains the pre-saccadic spike followed by a sharp monotonic increase (or decrease for look down and look left), and then there is a slow decay back to zero due to the high pass filter used in the signal acquisition hardware (Thickbroom & Mastaglia, 1986).

The raw EOG also contains micro-saccades, which unlike the major saccades are very small in amplitude but occur very frequently. These micro-saccades can occur in the middle of a major saccade (Figure 10.3). When this happens, the micro-saccades can cause the dynamic linear fit portion of the algorithm to be less accurate. Specifically, the full amplitude of the major saccade may not be detected. To minimize the effects of micro-saccades, the raw EOG data are filtered using a first order Butterworth low pass filter with a break frequency of 50 Hz.

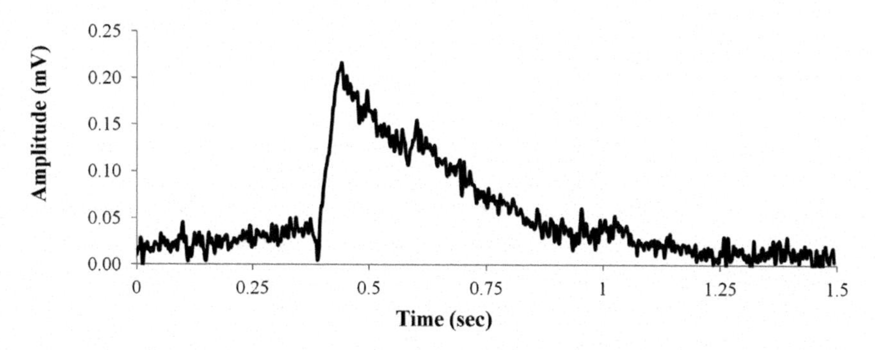

Figure 10.2 The typical shape of a saccade.

Note. This is a horizontal saccade to the right.

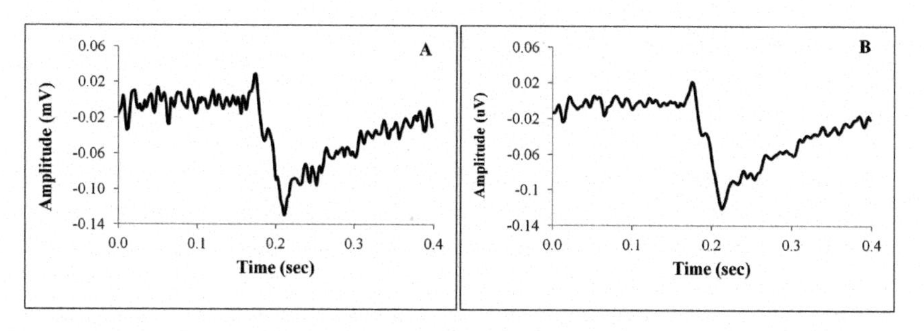

Figure 10.3 Horizontal saccade from a leftward eye movement.

It has a micro-saccade in the middle of it (A) that causes the linear fit to fall short of the full saccade amplitude. After filtering (B), the micro-saccade is reduced enough so that it does not interfere with the full linear fit.

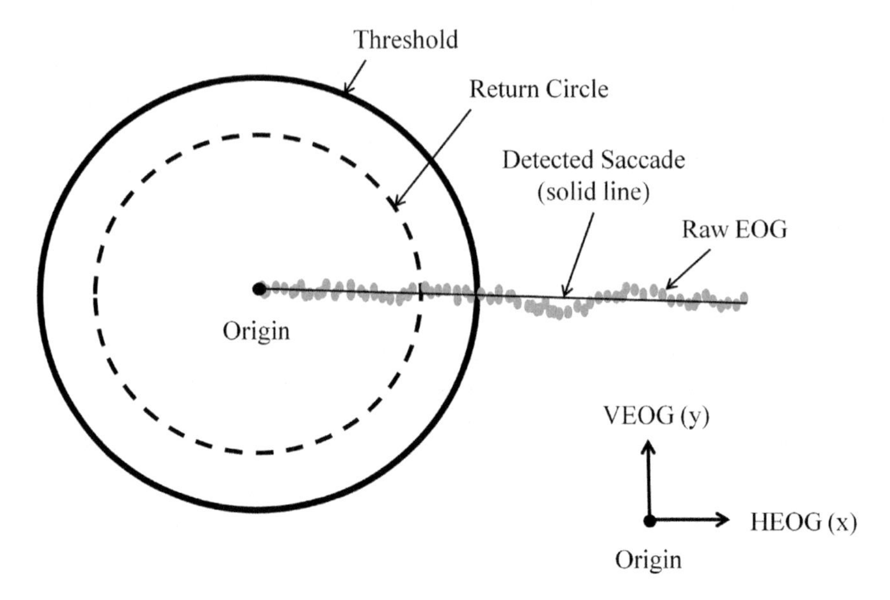

Figure 10.4 Polar saccade detection threshold, return circle, and raw signal.

Threshold determination. The threshold determination approach that was developed for the blink detection was adapted for the polar saccade detection algorithm with two distinctions. One, the polar saccade detection threshold does not incorporate the threshold reduction logic used for blink detection. Two, a circular threshold is used in the polar saccade detection algorithm such that saccades can be detected at any angle (Figure 10.4).

Saccade endpoint detection. The saccade endpoint detection is performed in Cartesian coordinates with HEOG on the x-axis and VEOG on the y-axis. The circular threshold is centered around the x/y origin. Initial saccade detection follows three simple steps. First, the x/y position must start inside the circular threshold. Second, the x/y position must travel outside the circular threshold. Third, the x/y position is allowed to move away from the origin for as many samples as it takes until it moves back toward the origin for two samples in a row. The last sample that is moving away from the origin is the endpoint of the saccade.

To minimize false positives, a second circle called the return circle was added. This circle is centered about the origin and its radius is equal to two-thirds of the threshold circle. The detection logic required that the signal must return to inside the return circle before it can be tested for traveling outside the threshold. The saccade endpoint detection is accomplished using a state machine.

Saccade start point determination. A dynamic linear fit was performed to detect the full amplitude of the saccade. The dynamic linear fit was performed in rectangular coordinates. That is, the VEOG and HEOG were processed separately based on the saccade endpoint. Two vectors were used to find the saccade starting point for each signal (VEOG and HEOG). These two vectors were referred to as the small vector and the big vector. The initial length of both vectors was 20 ms (Chen & Wise, 1996). The heads of the vectors were set to the saccade endpoint and the tails were 20 ms backwards in time. The length of the small vector remained constant but the big vector grew in length backwards in time. The tail of the small vector was anchored to the tail of the big vector. When the slope of the small vector differed substantially from the slope of the big vector, it marked the saccade starting point (Figure 10.5). After the dynamic linear fits were performed on both axes, the x and y coordinates of the starting point and ending point for the potential saccade are known.

Calculations. Once the coordinates of a potential saccade are known, the amplitude, length, velocity, and peak velocity can be calculated. First, the rectangular coordinates were converted to polar coordinates (magnitude and angle). For easier evaluation, the two R^2 values from the dynamic linear fits were combined into a single R^2 value. Also, fixation duration was computed by subtracting the times between saccades.

Saccade classification criteria. Consistent with the main saccade sequence, the classification component used three of the variables computed above (R^2, velocity, and length). These variables were compared to criteria values to determine if the potential saccade was an actual saccade. All three of these criteria had to be met for a positive saccade classification. The initial criteria values were determined using data from a mini-study with four participants. This was accomplished using an EOG playback feature. All of the potential saccades were hand-scored by trained observers to generate truth data. The initial values were refined when data from a follow-on study was available (Mead, Hoepf, Middendorf, & Gruenwald, in preparation). The final criteria values are shown in Table 10.3.

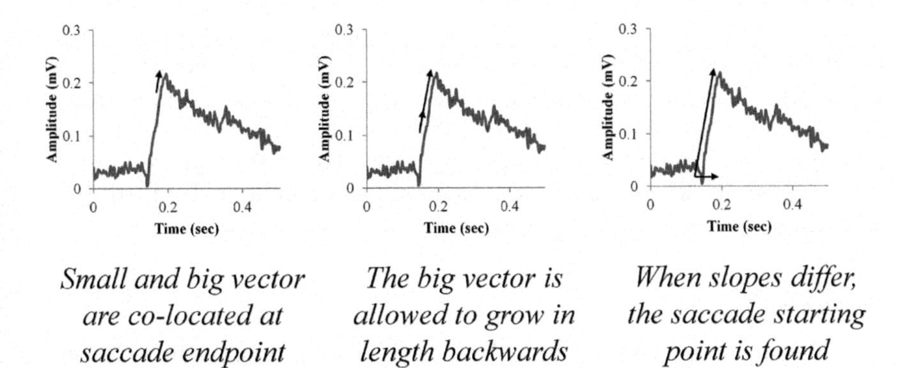

Figure 10.5 Linear fit vectors.

Table 10.3 Criteria values used for the saccade detection

Criteria	*Minimum value*	*Maximum value*
Combined R^2	0.85	N/A
Velocity (mV/s)	1.5	11.5
Length/duration (s)	0.028	0.125

Distinguishing saccades from blinks. A saccade queuing logic was needed to ensure that the potential saccade was not actually the up slope of a blink. While waiting for the excursion of the EOG signal to reach its maximum distance from the origin, the blink detection algorithm was monitored. If the blink detector was active, then a flag was set to indicate that the saccade must be queued. If a blink was detected, then the queued saccade was discarded, otherwise the queued saccade was counted. In essence, this means that blinks "trump" saccades. This was because the electrical signal for a blink is much larger than that for a saccade.

Initial validation test

For validation purposes, a study was conducted to test the performance of the polar saccade detection algorithm. In this study, visual stimuli were presented at known angles and distances at regular intervals (1.5 s). Two researchers independently reviewed the raw VEOG and HEOG signals to verify that the saccades were present in the signal. Following their independent reviews, they reconciled any difference by discussion. Compared to these signals classified by human raters, the saccade detection algorithm had 92.6 percent accuracy with zero false positives.

Algorithm calibration

To improve the accuracy of the algorithm's reported saccade angle and magnitude, software was developed to allow the two axes of EOG to be normalized, thus improving the accuracy of the reported saccade angles. The calibration process is needed because the VEOG signal is typically smaller than the HEOG signal for the same amount of angular movement of the eye. For some individuals, the difference in amplitude between the two axes can be as much as a factor of two. For others, the two axes have near symmetry. So, the calibration procedure can account for individual differences. The calibration procedure is easy to perform and only takes a few minutes. Results indicate that participant calibration is stable from day to day.

Summary

The measures produced by the polar saccade detection algorithm (velocity, amplitude, and length) have excellent precision. The saccade detection algorithm produced these measures in real time and in a computationally efficient manner. Another positive aspect of the current work is that it can support artifact mediation when performing EEG analysis.

An indirect use of the algorithms

The blink and saccade detection algorithms can be used directly as an indicator of increased cognitive workload. They also can be used indirectly in the analysis of EEG data to remove EOG artifacts, thereby improving the EEG measures' sensitivity to changes in mental workload. The presently developed algorithms were applied in a recent study reported below.

In the Hoepf et al. (in preparation) study described above, participants were outfitted with an electrode cap so that EEG data could be acquired. Seven channels of EEG data were recorded, including frontal sites F7 and F8. Power in the delta (1–3 Hz) band was computed for these sites. The VEOG data were acquired using two electrodes placed above and below the left eye.

Results showed that the route and number of targets manipulation had the expected effect on task performance and self-reported workload. EEG results revealed a significant effect of workload on frontal delta for the route manipulation. However, compared to the literature (Wang & Zhou, 2013), these findings were in the wrong direction. It was unclear if the observed frontal delta effects were due to brain activity or EOG artifacts. To investigate this concern, the blink and saccade detection algorithms were used to flag EEG spectral results. The flags were then used by the aforementioned artifact separation technique to ascertain the presence of EOG artifacts.

Before the artifact separation technique was applied to the EEG data (i.e., when all of the EEG frontal delta data were used without artifact separation), workload was indicated to be higher in the country than in the city condition at six sites. When blinks were separated, only two sites remained significant. When both blinks and saccades were separated, no increased workload was indicated in any of the sites. So removing the ocular artifacts addressed the concern about the frontal delta findings being in the wrong direction.

The results here showed that applying the artifact separation approach can change conclusions. One could reasonably argue that the artifact separation approach developed here is the same as the more common automatic artifact rejection approach. An important difference is that artifact rejection is typically done in the time domain and the artifact separation approach is done on the spectral results. Because the EEG spectral results are available in real time, they are available for the consumer of the data to decide what to do with the artifact flags. One practical use of the artifact flags is to use the flags to analyze only the contaminated data to solve problems associated with data collection (Credlebaugh, Middendorf, Middendorf, & Galster, in preparation). Another application is making use of the artifact flags in developing machine learning techniques in refining machine learning models.

Conclusion

Many researchers have tried to develop psychophysiological measures to improve the teamwork between human operators and systems that they must control. However, integrating the needed sensors and software was often too challenging for many individual research and development efforts. But with the recent breakthroughs in fairly nonintrusive physiological sensing capabilities, the viability of bringing together all of the necessary components should now be within reach.

The AFRL SAA program is designed to integrate sensors for physiological data acquisition, develop algorithms for feature extraction, assess the implications of the features, and inform augmentations that aid the human operators to achieve their missions. The current work demonstrated that relatively easily acquired eye activity data have great potential for the SAA framework. The sensing techniques and feature extraction algorithms needed for assessing eye data will be further refined. But more importantly, this approach will be extended to develop comparable processing of other physiological signals to produce robust assessments of the human operator's state. This will ultimately be used to implement appropriate augmentations.

References

Ahlstrom, U., & Friedman-Berg, F. J. (2006). Using eye movement activity as a correlate of cognitive workload. *International Journal of Industrial Ergonomics, 36*, 623–636.

Andreassi, J. L. (2007). *Psychophysiology: Human behavior and physiological response* (5th ed.). Mahwah, NJ: Erlbaum.

Arai, K., & Mardiyanto, R. (2010). Real time blinking detection based on Gabor Filters. *International Journal of Human Computer Interaction, 1*, 33–45.

Benedetto, S., Pedrotti, M., & Bridgeman, B. (2011). Microsaccades and exploratory saccades in a naturalistic environment. *Journal of Eye Movement Research, 4*, 1–10.

Bodala, I. P., Ke, Y., Mir, H., Thakor, N. V., & Al-Nashash, H. (2014). Cognitive workload estimation due to vague visual stimuli using saccadic eye movements. In *36th Annual International Conference of the IEEE Engineering in Medicine and Biology Society* (pp. 2993–2996). Chicago, IL: IEEE. doi: 10.1109/EMBC.2014.6944252.

Chen, L. L., & Wise, S. P. (1996). Evolution of directional preferences in the supplementary eye field during acquisition of conditional oculomotor associations. *The Journal of Neuroscience, 16*, 3067–3081.

Chen, S., & Epps, J. (2013). Automatic classification of eye activity for cognitive load measurement with emotion interference. *Computer Methods and Programs in Biomedicine, 110*, 111–124.

Credlebaugh, C., Middendorf, M., Hoepf, M., & Galster, S. (2015). EEG data analysis using artifact separation. In *Proceedings of the 18th International Symposium on Aviation Psychology* (pp. 434–439). Dayton, OH: International Symposium on Aviation Psychology.

Credlebaugh, C., Middendorf, P., Middendorf, M., & Galster, S. (in preparation). *An EEG data investigation using only artifacts* (Technical Report). Wright-Patterson Air Force Base, OH: United States Air Force Research Laboratory.

Di Stasi, L. L., Marchitto, M., Antolí, A., & Cañas, J. J. (2013). Saccadic peak velocity as an alternative index of operator attention: A short review. *Revue Européenne de Psychologie Appliquée, 63*, 335–343.

Di Stasi, L. L., Renner, R., Staehr, P., Helmert, J. R., Velichkovsky, B. M., Cañas, J. J., Cantena, A., & Pannasch, S. (2010). Saccadic peak velocity sensitivity to variations in mental workload. *Aviation, Space, and Environmental Medicine, 81*, 413–417.

Divjak, M., & Bischof, H. (2008). Real-time video-based eye blink analysis for detection of low blink-rate during computer use. In Gonzáles, J., Moeslund, T. B., & Wang, L. (eds.), *Tracking humans for the evaluation of their motion in image sequences, First International Workshop (THEMIS 2008)*, (pp. 99–107). Barcelona, Spain: Gràficas Rey, S. L.

Fatourechi, M., Bashashati, A., Ward, R. K., & Birch, G. E. (2007). EMG and EOG artifacts in brain computer interface systems: A survey. *Clinical Neurophysiology, 118*, 480–494.

Fitts, P. M., Jones, R. E., & Milton, J. L. (1950). Eye movements of aircraft pilots during instrument-landing approaches. *Aeronautical Engineering Review, 9*, 24–29.

Galster, S. M., & Johnson, E. M. (2013). *Sense-assess-augment: A taxonomy for human effectiveness* (Technical Report AFRL-RH-WP-TM-2013-0002). Wright-Patterson Air Force Base, OH: United States Air Force Research Laboratory.

Gao, Q., Wang, Y., Song, F., Li, Z., & Dong, X. (2013). Mental workload measurement for emergency operating procedures in digital nuclear power plants. *Ergonomics, 56*, 1070–1085.

Gevins, A. S., & Smith, M. E. (2003). Neurophysiological measures of cognitive workload during human-computer interaction. *Theoretical Issues in Ergonomics Science, 4*, 113–131.

Gill, A. (1962). *Introduction to the theory of finite-state machines*. New York: McGraw-Hill.

Gomer, F. E. (ed.) (1981). *Biocybernetic applications for military systems* (Report No. MDC E2191). Saint Louis, MO: McDonnell Douglas Astronautics Company—St. Louis Division.

Halverson, T., Estepp, J., Christensen, J., & Monnin, J. (2012). Classifying workload with eye movements in a complex task. In *Proceedings of the Human Factors and Ergonomics Society 56th Annual Meeting, 56*, 168–172. doi: 10.1177/1071181312561012.

Hankins, T. C., & Wilson, G. F. (1998). A comparison of heart rate, eye activity, EEG and subjective measures of pilot mental workload during flight. *Aviation, Space, and Environment Medicine, 69*, 360–367.

Hoepf, M., Middendorf, M., Epling, S., & Galster, S. (2015). Physiological indicators of workload in a remotely piloted aircraft simulation. In *Proceedings of the 18th International Symposium on Aviation Psychology* (pp. 428–433). Dayton, OH: International Symposium on Aviation Psychology.

Hoepf, M., Middendorf, M., & Galster, S. (in preparation). *Physiological indicators of workload in a surveillance and tracking task using remotely piloted aircraft* (Technical Report). Wright-Patterson Air Force Base, OH: United States Air Force Research Laboratory.

Jiang, X., Tien, G., Huang, D., Zheng, B., & Atkins, M. S. (2013). Capturing and evaluating blinks from video-based eyetrackers. *Behavior Research Methods, 45*, 656–663.

Kong, X., & Wilson, G. F. (1998). A new EOG-based eye-blink detection algorithm. *Behavior Research Methods, Instruments, & Computers, 30*, 713–719.

Kramer, A. F. (1991). Physiological metrics of mental workload: A review of recent progress. In D. L. Damos (ed.), *Multiple-task performance* (pp. 279–328). Bristol, PA: Taylor & Francis.

Marshall, S. P. (2007). Identifying cognitive state from eye metrics. *Aviation, Space, and Environmental Medicine, 78*, B165–B175.

May, J. G., Kennedy, R. S., Williams, M. C., Dunlap, W. P. & Brannan, J. R. (1986). *Eye movements as an index of workload* (Technical Report AFOSR-TR-86-0416). Bolling AFB, DC: Air Force Office of Scientific Research.

Mead, J., Hoepf, M., Middendorf, M., & Gruenwald, C. (in preparation). *Polar saccade detection algorithm: Classification criteria refinement* (Technical Report). Wright-Patterson Air Force Base, OH: United States Air Force Research Laboratory.

Parasuraman, R., & Galster, S. (2013). Sensing, assessing, and augmenting threat detection: Behavioral, neuroimaging, and brain stimulation evidence for the critical role of attention. *Frontiers in Human Neuroscience*, 7, Article 273. doi: 10.3389/fnhum.2013.00273.

Schmorrow, D. D. & Reeves, L. M (eds.) (2007). *Foundations of augmented cognition: Lecture notes in artificial intelligence*, 4565. Germany: Springer.

Sirevaag, E., Kramer, A., De Jong, R., & Mecklinger, A. (1988). A psychophysiological analysis of multi-task processing demands. *Psychophysiology, 25*, 482.

Stanney, K., Winslow, B., Hale, K., & Schmorrow, D. (2015). Augmented cognition. In D. A. Boehm-Davis, F. T. Durso, & J. D. Lee (eds.), *APA handbook of human systems integration* (pp. 329–343). Washington, DC: American Psychological Association.

Stern, J. A., & Skelly, J. J. (1984). The eyeblink and workload considerations. In *Proceedings of the Human Factors and Ergonomics Society 28th Annual Meeting* (pp. 942–944). Santa Monica, CA: Human Factors and Ergonomics Society.

Stern, J. A., Walrath, L. C., & Goldstein, R. (1984). The endogenous eyeblink. *Psychophysiology, 21*, 22–33.

Sukno, F. M., Pavani, S.-K., Butakoff, C., & Frangi, A. F. (2009). Automatic assessment of eye blinking patterns through statistical shape models. In M. Fritz, B. Schiele, & J. H. Piater (eds.), *Lecture notes in computer science, volume 5815* (pp. 33–42). New York: Springer.

Thickbroom, G. W., & Mastaglia F. L. (1986). Presaccadic spike potential. Relation to eye movement direction. *Electroencephalography and Clinical Neurophysiology, 64*, 211–214.

Toivanen, M., Pettersson, K., & Lukander, K. (2015). A probabilistic real-time algorithm for detecting blinks, saccades, and fixations from EOG data. *Journal of Eye Movement Research, 8*(2), 1–14.

Tokuda, S., Palmer, E., Merkle, E., & Chaparro, A. (2009). Using saccadic intrusions to quantify mental workload. In *Proceedings of the Human Factors and Ergonomics Society 53rd Annual Meeting* (pp. 809–813). Santa Monica, CA: Human Factors and Ergonomics Society.

Wang, Y., & Zhou, J. (2013). Attachment A: Literature review on physiological measures of cognitive workload. In F. Chen (ed.), *Robust multimodal cognitive load measurement (RMCLM)*. Report Number AOARD-124049. Sydney, Australia: National ICT Australia (NICTA).

Wickens, C. D., Hollands, J. G., Banbury, S., & Parasuraman, R. (2013). *Engineering psychology and human performance* (4th ed.). Boston, MA: Pearson.

Wierwille, W., Rahimi, M., & Casali, J. (1985). Evaluation of 20 workload measures using a simulated flight task emphasizing mediational activity. *Human Factors, 27*, 489–502.

Yang, J. H., Kennedy, Q., Sullivan, J., & Fricker, R. D., Jr. (2013). Pilot performance: Assessing how scan patterns & navigational assessments vary by flight experience. *Aviation, Space, and Environmental Medicine, 84*, 116–124.

11 Assessing situation awareness in an unmanned vehicle control task

A case for eye tracking based metrics

Joseph T. Coyne, Ciara M. Sibley, and Samuel S. Monfort

The promise of measuring situation awareness (SA) unobtrusively is a goal for many basic and applied researchers. Measurement in general works best when its administration does not affect its outcome. Nonetheless, many researchers have begun to view inadvertent measurement effects as a necessary cost, assessing SA by means of potentially disruptive performance probes and/or queries (e.g., Situation Awareness Global Assessment Technique (SAGAT), Situation Present Assessment Method (SPAM), etc.). Advances in eye tracking technology, however, have begun to facilitate continuous, unobtrusive inferences about underlying mental states, including workload (e.g., Ahlstrom & Friedman-Berg, 2006) and SA (e.g., Ratwani, McCurry, & Trafton, 2010), among others. Although these techniques have been employed by many researchers, certain methodological and procedural issues remain pertaining to how eye gaze data are processed and applied to inform human factors research. The current chapter aims to provide guidance on these issues by reviewing literature relevant to SA measurement and by presenting original empirical work on eye tracking data collected within a dynamic, complex Unmanned Aerial Vehicle (UAV) management simulation. The chapter discusses the strengths and limitations of the different SA measurement techniques and demonstrates how eye-based measures of SA can complement the existing measures.

Current and future challenges of unmanned aerial vehicle operation

The past decade has seen a dramatic rise in the use of UAVs within the US military. These systems accounted for only 5 percent of the Department of Defense's aircraft inventory in 2005, but by 2012 that number had surged to 41 percent (Gertler, 2012). Initially, UAVs provided niche capabilities to military operations, but this is no longer the case as the number, variety, and capabilities of these systems have rapidly expanded (Department of Defense, 2013). Furthermore, success within military operations has led to the exploration of UAV use within new civilian applications, from entertainment to security to drone delivery services.

The use of the word "unmanned" to describe these systems is clearly a misnomer. Although there is no pilot physically onboard the platforms, current unmanned vehicles are heavily reliant on human operators. In fact, the personnel on many unmanned systems exceeds that of manned aircraft with similar capabilities. For example, even small tactical UAVs may require up to three operators: a pilot who controls the vehicle, a payload operator who controls the sensor (typically a camera), and a mission commander who oversees the goals of the mission by directing the team and handling external communication. The rapid increase in UAV acquisition and the associated manpower costs has generated interest in shifting away from a direct control paradigm toward a supervisory control model where a single operator monitors multiple vehicles (Department of Defense, 2013). The majority of an operator's tasking under such a control model would involve assigning vehicles to pursue different targets and objectives, monitoring the progress of those vehicles, and updating plans given new opportunities and changing information. The responsibilities of this future operator will have drastically increased in scope and as such, will likely be associated with a number of human factors challenges. The challenge we address in this chapter stems from the difficulty associated with monitoring and accurately diagnosing lapses in SA. Continuous monitoring of operator SA represents a potential avenue for targeted interventions to improve engagement (Berka et al., 2007), operational efficiency (Endsley, 1999), and performance (Kohlmorgen et al., 2007).

Research has shown that humans are not well suited to monitoring complex automated systems and that there is a cost associated with human interaction that is limited to a supervisory capacity (e.g., Parasuraman, 1987). A number of classic human factors studies have revealed vigilance decrements in tasks when humans are required to monitor systems and identify rare events or system anomalies (Mackworth, 1948; Teichner, 1974; Warm, Parasuraman, & Matthews, 2008). The vigilance decrement often occurs when working with highly reliable automated systems, as humans become complacent and place too much trust in the automation (Parasuraman, Molloy, & Singh, 1993). The result of excessive trust is that humans shift their attention to other tasks and miss the rare automation failures when they do occur (i.e., automation "misuse"; Parasuraman & Riley, 1997). For example, researchers at the Air Force Research Laboratory found that after several sessions in a UAV supervisory control task wherein automation performed perfectly, every participant missed a route error introduced by the automation, despite being told that it could make mistakes (Calhoun, Draper, & Ruff, 2009). The shift in attention away from automation monitoring results in a decrease in the operator's SA, which can cause errors or hinder performance when a critical event occurs.

SA theory and measurement

The most widely used definition of SA from Endsley (1988a) is "the perception of elements in the environment within a volume of time and space, the comprehension of their meaning, and the projection of their status in the near future" (Endsley, 1988a, p. 97). While some researchers differ on whether SA refers to the process of gathering information (e.g., Adams, Tenney, & Pew, 1995) or to the product of gathering information (Endsley, 1995), the construct has meaning and value to the domains where it is applied regardless as to how one defines it. Thus, the practical issue that emerges is how to assess SA. As with other constructs that evade straightforward measurement, such as mental workload, there is a variety of metrics available for assessing SA, including freeze probes, in-task queries, subjective measurements, and performance measures. We review these methods below.

Freeze probes, such as the SAGAT (Endsley, 1988b), are some of the most common methods utilized for SA assessment. These probes are used in simulation environments where the task can be paused, during which time the interface is hidden and the individual is asked a specific series of questions regarding the current state of the environment. Probes are advantageous because they provide a direct measurement of SA; however, they have been criticized for being disruptive and for only providing data at discrete intervals. They are also limited in the sense that they can only be used within controlled laboratory environments (Salmon, Stanton, Walker, & Green, 2006).

One alternative to freeze probe measures are in-task queries in which the operator is queried while the scenario is still active. The SPAM (Durso & Dattel, 2004), for example, informs an operator that a query is in his or her queue, after which the operator waits to be available and then signals that he or she is ready to respond. The time required to respond to the query and the response accuracy are both considered as measures of SA, with faster, correct responses signifying a better understanding of the current state. These in-task queries are well suited for real-world environments; however, the act of responding to the probe may actually direct attention to an element of the environment of which the individual was not aware.

Post-task subjective SA measures are also quite popular, such as the Situation Awareness Rating Technique (SART) (Taylor, 1990). The SART requires participants to rate (from 1 = Low to 7 = High) the following ten dimensions after task completion: familiarity of the situation, focusing of attention, information quantity, information quality, instability of the situation, concentration of attention, complexity of the situation, variability of the situation, arousal, and spare mental capacity. A shorter, three-dimension SART assessment is also employed which only assesses attentional demand, attentional supply, and understanding. The use of subjective rating scales to assess SA is popular because they are easy to administer and are non-intrusive, two properties which make them good candidates for use in real-world environments in which tasks cannot be interrupted (Jones, 2000). However, subjective SA probes only provide insight into how aware the individual felt he or she was during the task. His or her subjective assessment may deviate significantly from "actual" SA (Salmon et al., 2006). Past research supports these concerns, finding that SART scores were uncorrelated with SAGAT scores (Salmon et al., 2009). Self-rating of SA may only measure participant confidence in SA, rather than SA itself (Endsley, 1995).

Another method of assessing SA is to infer it from an individual's task performance. While good performance does not necessarily equate to good SA, it follows that an operator who is more aware of their surroundings will be better able to respond to unexpected challenges, should they occur. For example, a UAV operator aware of deteriorating weather conditions might be more likely to alter course than one who is unaware of the impending weather. The correspondence between SA and performance will not be perfect, however. An operator might be unaware of the weather conditions, but by chance not be adversely impacted by the weather, resulting in high performance despite low SA. Individual differences in skill may also cloud the relationship between SA and performance, as more talented pilots may be better able to infer overall system status from limited information (Lathan & Tracey, 2002). Lastly, using operator performance to assess SA would be impractical in situations where the performance outcomes are limited: monitoring a location of interest while no activity occurs, for example. Thus performance monitoring represents an important, yet limited method for SA assessment.

Eye tracking and SA

Traditional measures of SA (i.e., subjective and behavioral) represent a useful but nonetheless incomplete avenue for inferring UAV operator attentional state. A number of recent studies (Moore & Gugerty, 2010; Ratwani et al., 2010; Van de Merwe, Van Dijk, & Zon, 2012; Gartenberg, Breslow, McCurry, & Trafton, 2014) have examined the use of eye tracking as a means of supplementing existing SA assessments. Eye tracking has a number of obvious benefits, such as providing continuous objective data without interrupting tasking (Van de Merwe et al., 2012). Additionally, while other measures of SA focus on SA as an outcome (a knowledge product; Endsley, 1995), eye tracking may also provide insight into the process of acquiring and maintaining SA (Adams et al., 1995). The addition of eye tracking methods to existing SA assessment protocols may provide a more complete understanding of a user's SA and also alleviate some of the shortcomings associated with traditional assessments.

Eye tracking provides researchers with a means of continuously and unobtrusively assessing where an individual is allocating his or her visual attention. Although there are a few situations where an individual may distribute attention elsewhere from his or her eye gaze location, the connection between gaze position and attention is well-established (Rayner, 2009). Employed during reading comprehension tasks, eye tracking can provide a wealth of information on information processing, indicating which words require additional processing and which words are skipped (Rayner, 2009). There has been an abundance of research on eye tracking during simple tasks such as reading, where movements are serial and can be modeled, but the use of eye tracking in more complex visual environments is less advanced.

In complex environments, such as a cockpit or automobile, an individual's attention (i.e., gaze pattern) is influenced by a number of factors, including the individual's experience level and the specific situation (Bellenkes, Wickens, & Kramer, 1997; Crundall, Underwood, & Chapman, 1999; Schriver, Morrow, Wickens, & Talleur, 2008). Theories of expertise have shown that experts can quickly recognize patterns in the environment and assess a situation (e.g., Ericsson & Kintsch, 1995; Klein, 1997), which has led to a number of researchers using eye tracking as a way of learning where experts focus their attention. One of the primary means of categorizing visual focus involves discretizing the visual area into several areas of interest (AOIs) based upon those areas' functions and the specific research questions (Holmqvist et al., 2011).

Gegenfurtner et al. (2011) conducted a meta-analysis on sixty-five articles, utilizing eye tracking to investigate expertise across a wide variety of domain applications. Unsurprisingly, expertise was strongly correlated with performance across domains ($r = .49$). In addition, experts tend to have more fixations on task relevant AOIs ($r = .56$) and fewer fixations on redundant AOIs ($r = -.32$). These results suggest that experts' attention allocation strategies, as assessed through eye gaze, may account for some of the observed performance differences. For example, Schriver et al. (2008) found that expert pilots diagnosing an engine failure spend more time looking at relevant AOIs than do non-expert pilots. Additionally, when multiple cues were provided that varied in their diagnostic value, expert pilots spent an increased amount of time looking at the highly diagnostic AOIs while the less experienced pilots spent more time looking at the less diagnostic AOIs. Thus the length and target of visual fixations seems to be involved with the process of acquiring SA. The role of visual fixations in SA acquisition is also supported by Ratwani et al. (2010), who found a similar effect in a UAV management simulation (Boussemart & Cummings, 2008). Lastly, Gartenberg et al. (2014) found that the number of visual fixations after an interruption increased and the average fixation duration decreased, indicating that series of short, high-frequency fixations might be indicative of an individual's attempt to regain SA. In sum, understanding how to allocate attention can be considered a byproduct of and a precursor to good SA, and eye tracking represents a means for measuring this understanding.

Although research has provided compelling evidence for the use of eye tracking as a predictor of SA, the experimental design typically employed by this research tends to be somewhat artificial. For example, Ratwani et al.'s (2010) use of a UAV simulation paradigm to assess operator SA presented high-frequency events that required frequent monitoring by the individuals. Periods of high-frequency changes are atypical of interactions with highly automated UAVs, which instead tend to contain periods of very low workload within a static environment. Indeed, UAV operators tend to struggle with boredom more than with overload (Cummings, Mastracchio, Thornburg, & Mkrtchyan, 2013).

There have been a number of studies which utilize eye tracking in high-fidelity, realistic environments. Moore and Gugerty (2010) compared eye gaze measures of SA and probe-based SA measures within a realistic air traffic control simulation of Los Angeles airport. The controllers completed three fifteen-minute scenarios with an SA freeze probe which asked about aircraft they were monitoring, occurring around eight minutes into each scenario. The authors collected eye tracking data throughout the simulation and their analysis focused on: percent time fixating on aircraft, standard deviation of percent time focusing on aircraft, mean time fixating on the aircraft, total number of fixations, and a measure of fixation dispersion known as the nearest neighbor index (NNI). Of these metrics, only the percent and standard deviation of time fixating on aircraft successfully predicted performance on the SA probes. However, an increased NNI, which indicates a more dispersed gaze pattern, was predictive of the number of errors made during the simulation.

Research by Van de Merwe, Van Dijk, and Zon (2012) investigated the use of eye tracking during a simulated flight using teams of commercial airline pilots. During the simulated flight, pilots were alerted to an "undefined system mismatch" via the aircraft display. After notification, the pilots had five minutes to determine the cause of the malfunction. The researchers found that the pilots who identified the problem increased the amount of time spent looking at the warning system by 45 percent compared with their pre-malfunction period, whereas pilots who failed to identify the problem only increased the time looking at the warning system by 8 percent. Additionally, the authors found that the scan patterns of the group who failed to identify the problem became more dispersed, as assessed via the NNI, compared to the scan patterns of those who found the source of the problem. This suggested that the group that could not identify the problem did not know where to direct their attention. In sum, recent research suggests that eye tracking is a promising metric for assessing operator behavior and inferring underlying cognitive states, even in complex, high-fidelity simulations.

Eye tracking measures of SA

Within the eye tracking literature, eye movements are divided into two categories: saccades, which are periods of rapid eye movements, and fixations, which are the period of time in which the eye remains relatively stable. Visual input is suppressed during the saccade, so it is during fixations that an individual processes information about the visual scene (Rayner, 2009). There are two main types of fixation detection algorithms: dispersion-based and velocity-based. Within the dispersion category, the primary approaches for defining fixation area entail either calculating the maximum distance from a center point of raw data points (centroid), or calculating distance based upon the furthest x and y-points within a group of raw data points. A series of consecutive samples that are sufficiently close to each other are coded as a fixation when the number of samples meets the specified minimum time duration. The maximum distance and duration are specified by the researcher. The velocity-based approach computes fixations based upon the angular velocity between consecutive points, typically defined as degrees per second. The velocity approach does not necessarily have a minimum time duration associated with it. See Holmqvist et al. (2011) for a detailed discussion of the various techniques.

Although there is active debate over the different approaches and thresholds for defining a fixation (Collewijn & Kowler, 2008; Holmqvist et al., 2011), many researchers may not be aware of the variety of techniques and rely on their eye tracking software to select a method to compute fixations. Indeed, all too often researchers fail to report the specific techniques or thresholds employed to calculate fixations, and even those who do cannot rely on standardized guidelines because they do not yet exist. The studies discussed previously are unfortunately a good representation of the variety of thresholds used and lack of complete reporting methods. Specifically, of the four studies described above, three different minimum fixation durations were used: 50 ms (Ratwani et al., 2010), 100 ms (Moore & Gugerty, 2010) and 150 ms (Van de Merwe et al., 2012); one did not report a minimum duration (Gartenberg et al., 2014). Furthermore, the reported maximum distance thresholds varied from one to two degrees of visual angle, and one reported using 30 pixels but failed to provide the distance to the monitor so it could not be converted to visual angle. Specifying the visual angle in eye tracking research is important since it allows comparison of results across studies with different viewing distances, monitor sizes and monitor pixel densities. Furthermore, two of the four studies presented were unclear on the algorithms used to compute fixations, and of the remaining two, one used a velocity algorithm and the other used a centroid dispersion algorithm.

Research suggests that the algorithms and thresholds used to characterize visual fixations can greatly influence the results of a study. Thus, future papers using fixation data may want to report the results for different thresholds (Shic, Scassellati, & Chawarska, 2008). Regardless, the lack of a standard definition for a fixation limits the degree to which eye tracking results can be compared across studies and makes understanding the field through meta-analysis impossible. Inconsistent fixation calculations not only impact the fixation measure itself, but also other metrics which are derived from fixations, such as the NNI, which is a measure of dispersion (Clark & Evans, 1954) and entropy, which is a measure of the randomness of gaze patterns (Pincus, 1991).

The NNI is the ratio between the average reported minimum distances between fixations and the mean distance that one would expect if the distribution were random (Clark & Evans, 1954). The NNI was originally used in reference to plant and animal populations, and was first used by Di Nocera, Terenzi, and Camilli (2006) as a means of comparing distributions of eye fixation data. The NNI is computed by dividing the averaged nearest neighbor distance (distance between each fixation and the next closest fixation) by the mean random distance (computed using the area fixations occur within and the number of points). Values for the NNI metric can range from 0 to 2.1491, where 0 represents maximum clustering, 1 is random, and values greater than 1 become increasingly evenly dispersed. Originally, Di Nocera and colleagues applied NNI as a measure of operator workload. They found that as workload increased during an asteroids game, the NNI suggested a more random gaze dispersion. The authors found a similar pattern with increasing NNI during more difficult phases of flight (Di Nocera, Camilli, & Terenzi, 2007). Thus, for tasks that are facilitated by dispersed visual sampling, individuals seem to adopt a strategy of dispersing their fixations, and the NNI metric is able to capture this dispersion.

Although workload and SA are different constructs, the two are related and perhaps causally connected. For example, a busy operator may be less able to completely process and make sense of information from his or her surroundings. Moore and Gugerty (2010) found that higher NNI was predictive of errors in an air traffic control task and as noted above, Van de Merwe et al. (2012) observed increased NNI values for all participants following the alert compared to before the alert. Additionally, the group who failed to identify the problem source had higher NNI values compared to those who did identify the problem, possibly indicating poor SA. These studies suggest that an increasingly dispersed or random fixation pattern may suggest a lack of SA, at least when coupled with tasks in which participants had to determine the cause of a problem. In sum, NNI and other metrics derived from fixation data seem to be a promising but underutilized metric for assessing SA.

Supervisory control operations user testbed (SCOUT™)

We developed a multiple-UAV management simulation to facilitate the study of eye tracking data in a dynamic, complex environment. Before presenting an empirical study with this paradigm, we briefly describe the supervisory control operations user testbed (SCOUT™) and its capabilities.

SCOUT was developed by the Naval Research Laboratory (NRL) to investigate research questions surrounding human-automation interaction with respect to the management of multiple UAVs. As such, the missions within SCOUT involve three heterogeneous UAVs and are representative of future tasking that UAV operators may be required to conduct. Tasks within SCOUT were developed through observation and interaction with current UAV operators and involved an iterative design and feedback process. These tasks primarily include mission planning (developing routes based on mission goals); airspace management (requesting access to controlled airspace); communication (responding to requests for information); and updating/adjusting air vehicle parameters (e.g., altitude and speed) in addition to target parameters (e.g., location and search radius) as new information is provided or commands are issued. Figure 11.1 depicts SCOUT's dual screen game environment setup.

SCOUT was developed to be game-like in order to inspire competition and to encourage user motivation. Consequently, users are given points for successfully accomplishing tasks and are shown their total mission score throughout game-play. The mission planning task is the primary source of points and requires the user to determine the best routes for sending each UAV to search

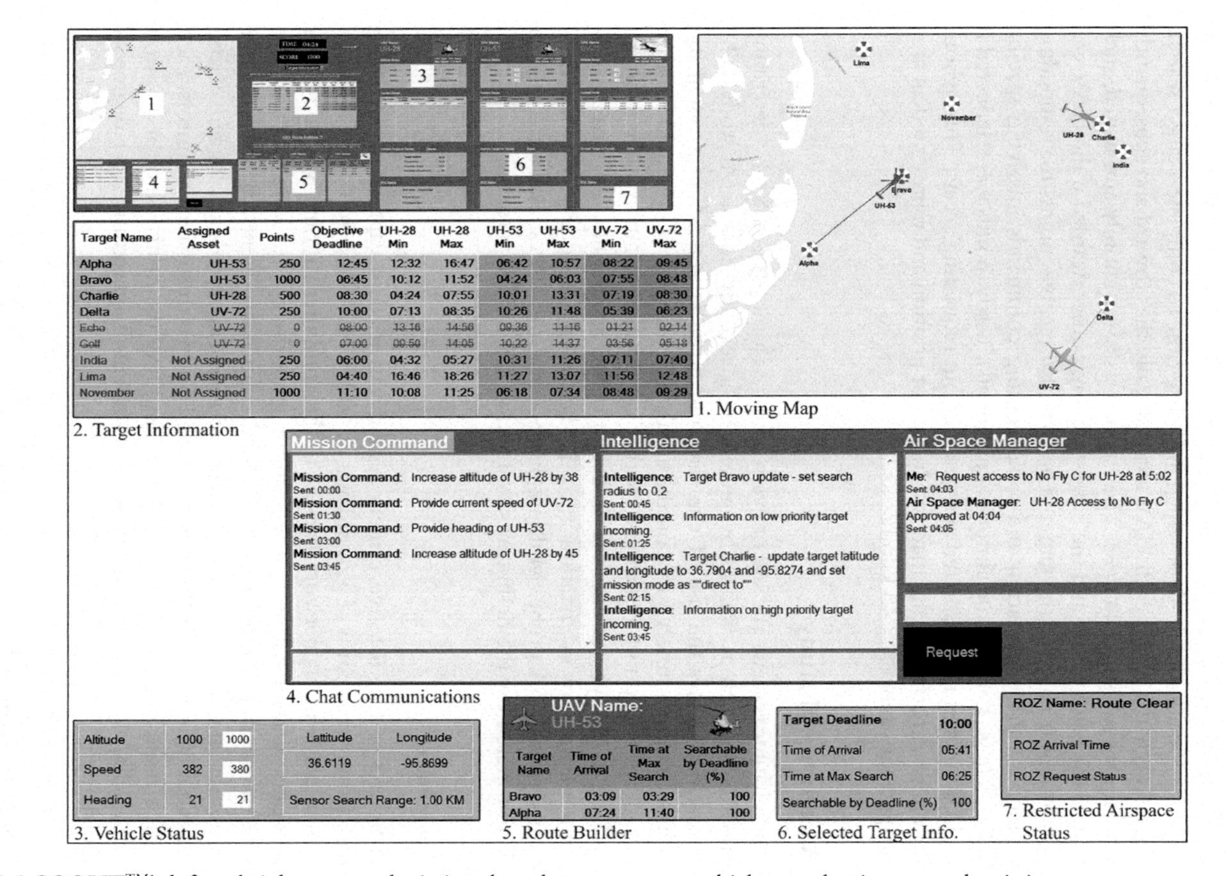

Figure 11.1 SCOUT™'s left and right screens depicting three heterogeneous vehicles conducting a search mission.

for targets with varying priority levels (point values), deadlines, and location uncertainties (influencing potential time required to locate the target). At the beginning of each mission, seven targets are available to pursue, but new targets appear throughout game-play. The appearance of new targets may cause the user to reconsider his or her plan. The three UAVs also differ in speed and sensor range capabilities, both of which influence time to complete a target search. Once assigned a route, each UAV automatically flies toward and subsequently searches for its assigned target. The payload task was entirely automated, such that a target was found and the corresponding points awarded if the UAV was within sensor range of the target. The user also receives points for responding to requests for information and for executing commands involving specific vehicles (e.g., heading and altitude adjustments). Lastly, points were deducted if a UAV entered into restricted airspace without requesting access. Throughout the entire experiment, the map was locked in position and represented an area of approximately 65 × 45 kilometers.

An empirical study

This chapter reviews the use of eye tracking as a measure of SA by describing previous research as well as by presenting a recently completed study. The research examines the use of eye tracking measures and SA in a systematic fashion within a supervisory control simulation. The results presented here are not meant to support a specific hypothesis, but rather to guide others who are looking to use eye tracking as an indicator of SA. Specifically, the current study demonstrates how eye fixations recorded before an SA probe vary with performance on that SA probe. Multiple distance and time thresholds were used to demonstrate how the definition of a fixation can influence results.

Equipment

A SmartEye Pro 6.1 five camera system was used to capture eye tracking data from participants at a frequency of 60 Hz. The data from the SmartEye system was sent via network packets to the computer running SCOUT and a module was built within SCOUT to integrate the eye tracking data with the simulation data. Additionally, SCOUT simulation events, behavioral/user response data, and eye tracking data were synchronized and logged within the same data files. SCOUT was displayed on two thirty-inch monitors, each with a resolution of 2560 × 1600. The width of the display and the position of the chair were such that 50 pixels represented approximately 1 degree of visual angle.

Experimental design

Twenty-three NRL civilian employees and summer interns with no prior operational UAV experience volunteered for participation in this experiment.

Three participants were excused from the study: two due to a lack of comprehension and one due to a simulation error. Analyses were conducted on the twenty remaining participants (seven women and thirteen men) ranging in age from 18 to 48 years ($M = 30$, $SD = 9.6$).

Participants completed approximately thirty minutes of training which consisted of a series of videos and sample tasks within SCOUT. If task comprehension was demonstrated through successful completion of the sample tasks, participants were invited to continue and engage in two experimental sessions, which were counterbalanced to mitigate order effects.

Each experimental session was pre-scripted and included four segments that always occurred in the following order: Planning, Block 1, Block 2, Block 3. During the Planning segment, UAVs were stationary and participants were given ten minutes to create an initial plan for sending their three UAVs to search for seven targets. After the planning segment, participants completed an eighteen-minute mission execution phase that was divided into three blocks of varying difficulty. Difficulty was manipulated via the frequency of chat tasking and the frequency of new targets added (see Table 11.1). Chat messages instructed the participant to update flight parameters on specific vehicles (e.g., "Decrease altitude of UH-28 by 117") and provided updated target information to help the user find targets more quickly (e.g., "Decrease search radius of target alpha to 1.5 km").

A freeze probe similar to the SAGAT was used to assess SA and was presented once within each six-minute block (three times per session, amounting to a total of six probes per participant). The computer-based probe used was designed to assess awareness of each vehicle and target's last position, the target each vehicle was traveling toward, and the point value of each target being pursued. During the SA probe, the SCOUT control displays would disappear and the probe would appear with the vehicles and targets randomly placed in the center of the map. At this time, the participant had two minutes to drag the vehicles and targets on the map to where they last recalled them being located. Participants then indicated which target each vehicle was currently pursuing, as well as that target's point value. These questions were intended to represent SA levels one and two, perception and comprehension (Endsley, 1995).

Once participants submitted their answers, they could assess their performance by viewing the correct locations of all items on the map, overlaid on their answers. Additionally, they were provided with the kilometer difference in distance from where they placed an item compared to where it actually was

Table 11.1 Experimental manipulation of difficulty

Block difficulty level	*Chat task frequency*	*New targets added*
Easy	75 seconds	1
Medium	45 seconds	3
Hard	15 seconds	4

before the probe. Participants received points for more accurate performance, indicated by lower distance errors and correct UAV-target pairs. After displaying the correct answers to probe questions, the simulation would return to its previous point and vehicles would only begin moving again once a "resume mission" button was pressed.

Results

SA probe results. The primary metric of the SA probe performance was the error distance between the participant's placement of UAVs and targets compared to the actual map position of those objects in the simulation immediately preceding the probe. Distance was computed in kilometers and the maximum possible error for each object was limited to 35 km so as to not skew results with large errors. A one-way repeated measures ANOVA revealed a block difficulty effect on the average SA probe distance error for targets and UAVs, $F(2,38) = 4.14$, $p = .024$. A Tukey HSD post hoc analysis revealed that average probe error was significantly lower in the easy ($M = 7.92$ km) difficulty block compared to the hard ($M = 10.69$ km) block, $p = .018$, while the medium block's error ($M = 9.18$ km) was not significantly different from either.

A secondary metric considered from the SA probe performance was the participant's ability to correctly identify which target each UAV was pursuing. The percentage of correctly assigned UAV-target pairs mirrored those of error distance, with a main effect of difficulty $F(2,38) = 7.826$, $p = .001$. A Tukey HSD post hoc analysis showed that the easy condition ($M = 61$ percent) had significantly better performance than the hard condition ($M = 36$ percent), $p = .001$.

Eye fixation analysis. The raw eye gaze data collected one minute before each SA probe were converted into fixations using both the centroid dispersion algorithm as well as the velocity-based algorithm. Both techniques were used to compare whether their results differed in terms of their correspondence with SA probe performance. First, however, we provide details on how each of the algorithms was computed.

The centroid dispersion-based method gathers the minimum number of samples required for a fixation (duration threshold), computes a centroid for those samples, and then determines if all of the points were within an experimenter-specified radius of the centroid (i.e., the distance threshold). If the minimum number of points do not fall within the set distance, the first sample in the window is dropped and a new sample is added, and the process begins again. If the minimum number of samples were within the maximum distance radius of the centroid, a fixation was identified and the number of samples within the fixation expanded while the subsequent samples met the distance criteria. The centroid was recomputed with each new sample and all points were rechecked until a new sample was determined to be outside the distance threshold.

Velocity-based algorithms identify raw samples as either part of a fixation or a saccade based upon the angular velocity between successive packets. The typical application of a velocity-based algorithm does not include a minimum time duration to be met (Holmqvist et al., 2011), meaning just two raw samples could comprise a fixation if the velocity between those points is slow enough. However, researchers will sometimes apply time constraints to velocity-based fixation metrics (Shic et al., 2008) and some software packages such as Tobii's ClearView compute fixations with a velocity-based algorithm, but use both distance and duration settings. The use of maximum distance and minimum duration settings also enables comparison of results acquired from centroid dispersion-based and velocity-based approaches when settings are held constant. As such, the velocity-based approach used to analyze the data computed fixations in the same way as the ClearView algorithm; by calculating the distance between successive packets (this is essentially the same, as computing velocity since the time between successive samples is constant). Each sample was only compared to the sample immediately before and after it. When a minimum number of successive samples (as determined by the minimum duration setting) were within the specified distance, that group of samples was marked as a fixation. The fixation was extended until a new sample was not within the maximum distance of the sample preceding it.

The centroid dispersion-based method is more computationally complex and intensive than the velocity-based approach because it requires the computation of a new centroid with each new sample that is considered, and then a comparison of every sample to the recomputed centroid. In contrast, the velocity-based approach only compares each sample with the sample before and after it (Salvucci & Goldberg, 2000). Nonetheless, the data for the current study were analyzed post hoc using both the centroid dispersion-based method and a velocity-based method (employing an additional minimum time duration). Six different fixation thresholds were computed for each approach: a combination of three minimum time durations (50, 100, and 150 ms) and two maximum distances (50 and 100 pixels, corresponding to approximately 1 and 2 degrees of visual angle). This analysis was conducted to demonstrate the impact different fixation criteria can have on results.

To complete the SA probes, participants needed to have an awareness of the relative positions of their UAVs and targets on the map. Given the demonstrated link between visual fixations and SA acquisition (e.g., Schriver et al., 2008; Ratwani et al., 2010; Gartenberg et al., 2014), our analyses focused on fixations within the map space in the one minute preceding each SA probe. Figure 11.2 depicts the correlation between the number of fixations in the map (based upon the six different thresholds and two different techniques) and SA probe error distance. The threshold combinations that were significantly correlated with SA error distance in Figure 11.2 were also significantly correlated with the UAV-target pairing portion of the probe and are thus not shown.

The pattern of correlations depicted in Figure 11.2 demonstrate that the fixation threshold criteria or settings used (i.e., minimum duration and maximum distance) can impact results, both in terms of strength of associations and where significance is found. The results also suggest that these settings can impact

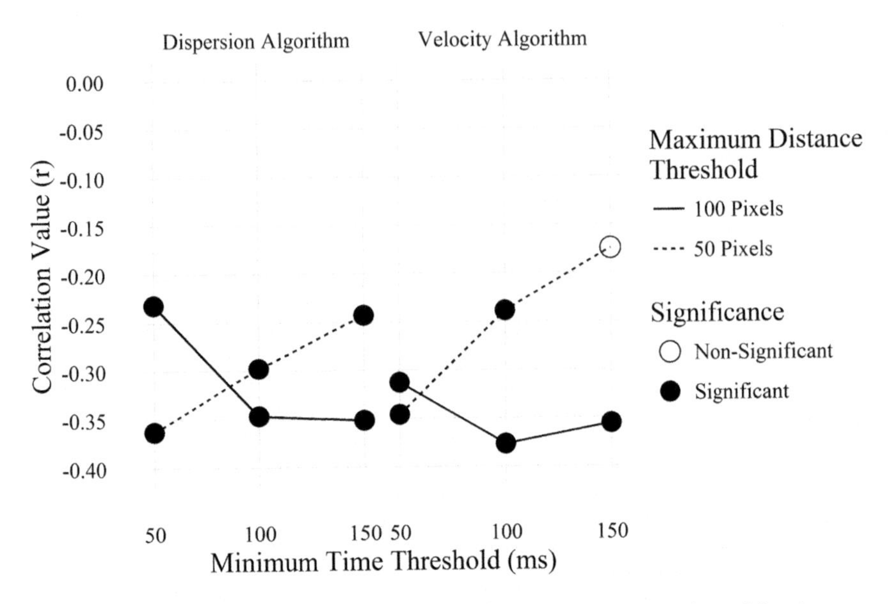

Figure 11.2 Correlations between SA probe error distance and number of fixations within the map using a velocity fixation algorithm and a centroid dispersion algorithm.

Note. The pattern of results suggest that the fixation thresholds can impact the results more than the type of algorithm selected.

results for both velocity-based and dispersion-based methods. All subsequent analyses were conducted using the velocity-based algorithm with a maximum distance threshold of 100 pixels (i.e., approximately 2 degrees of visual angle) and with a minimum time duration of 100 ms, since these settings resulted in the highest correlations between fixations and SA. The velocity-based algorithm also has the advantage of being computationally simpler (making it easier to implement in real time) than the centroid dispersion algorithm.

Eye fixation and difficulty. A one-way repeated measures ANOVA revealed a significant main effect of difficulty on the number of fixations in the map $F(2,38) = 35.59$, $p < .001$. A Tukey HSD post hoc analysis revealed that participants had significantly fewer fixations one minute prior to the SA probe for the hard block ($M = 13.18$) compared to both the easy ($M = 37.98$) and medium ($M = 27.00$) blocks. Similar results were found for the total time in seconds fixating within the map, where easy and medium blocks ($M = 12.09$ and $M = 8.65$, respectively) had significantly longer total fixation times than the hard block ($M = 3.96$).

NNI. The NNI was computed for each screen using the fixation data one minute before each SA probe. The NNI was only computed for blocks when at least five fixations were present. A one-way repeated measures ANOVA comparing NNIs across difficulty blocks showed no effect of difficulty (i.e., task load) on NNI values for the left screen. However, the right screen had significantly larger NNI ($M = 1.02$) in the easy condition than in hard ($M = 0.83$) conditions, denoting more random pattern of fixations as difficulty decreased $F(2,38) = 4.58$, $p = .018$. The NNI values were not significantly correlated with either SA error distance or UAV-target pair accuracy for either the left or right screens.

Discussion

Application of eye-based metrics of SA. Traditional measures of SA, such as SAGAT and the freeze probe within SCOUT, emphasize SA as a knowledge product while eye tracking provides insight into the process used to acquire SA. The results presented in this chapter are consistent with other findings (e.g., Moore & Gugerty, 2010) and provide further evidence that eye-based measures can inform understanding about an individual's SA.

One of the primary benefits of eye tracking over data generated from freeze probes is the continuous data stream it can provide. The true value of eye tracking based measures is not predicting SA performance on a static probe, but rather serving as a measure of an individual's dynamic SA throughout a task. The fact that just one minute of fixation data collected during our highly dynamic task could predict subsequent SA probe performance suggests that these measures can indeed serve as an online measure of SA. We therefore contend that this is a research area ripe for further investigation.

The present data also demonstrate how the specific thresholds used to define a fixation can influence a study's results and are intended to serve as an advisory to other researchers using gaze-based measures. Although other researchers have compared a number of different fixation algorithms (e.g., Salvucci & Goldberg, 2000), our results suggest that the choice of which fixation algorithm to use may be less critical than the choice of thresholds to apply with that algorithm. By adjusting the fixation distance/time thresholds, we showed that some of the more restrictive fixation criteria (e.g., 50 pixels and 150 ms) impacted results, such that the number of fixations within the map was no longer correlated with performance on the subsequent SA probe. Given the inconsistency of fixation thresholds applied within the human factors community, it is important for researchers to consider multiple thresholds and not simply assume the default settings used by their eye tracking systems will best suit their study's needs.

The use of NNI to characterize gaze patterns is still a relatively new practice, and the current study provided further evidence for the usefulness of NNI as a predictor of workload. However, similar to Moore and Gugerty's (2010) work, the present NNI data did not correlate with performance on an SA probe. Although a link between NNI and probe-based SA measures has not yet been established, results from Van de Merwe et al. (2012) suggested that pilots who were unable to identify the cause of a failure (less SA) had an NNI which became increasingly random (closer to 1) compared to those who were able to identify the problem. This may suggest that NNI could help identify SA in situations where an individual is responding to an unusual event. Specifically, a more dispersed gaze pattern may suggest that the individual does not know where to allocate his attention. One of the advantages to the NNI measure is that it does not rely on predefined areas on interest, but rather just looks at the how fixations are dispersed, so it would be valuable to explore its use and links with SA more thoroughly.

Limitations. One of the limitations of the data presented here is that the map was treated as a single AOI. Although the number of fixations on the map was predictive of SA performance, treating the map as a single AOI prevented a finer-grained analysis. An alternative approach would be to treat specific vehicles and targets as their own AOIs. This functionality has been added to SCOUT and future studies using this testbed will be able to correlate SA for each object with fixations on more precise regions.

Another limitation of the current study, and remote eye tracking studies in general, is that large head and body movements can result in lost data. Given that the time window examined was relatively long (one minute) and the eye tracker was capturing data at a high frequency (60 Hz), this meant that there were always data for each participant, but variable gaps occurred. We have since added a data alert to the SCOUT testbed which provides a visual warning to the participant if prolonged periods of bad data (which were often the result of slouching) are detected. Nonetheless, the visual alert requires testing: it has the potential to improve the quality of the eye tracking data but may also impact the task performance and scan patterns.

Although not necessarily a limitation, it is important to note that within the present study task demands were driven by chat communication as well as the inclusion of new targets. These tasks often took the operator's attention away from the map display. The eye movement data showed that as time spent looking at the map increased, overall SA probe error tended to decrease. It is unclear how queries or tasking that drew attention toward, rather than away from, the map display would have impacted either the eye tracking metrics or the performance on the SA probe.

Future research. The use of eye gaze-based measures of SA are still in the early stages of research, but the development of new low-cost ($100–$500) eye tracking systems such as Eye Tribe, Tobii's Eye X, and Gazepoint are making eye tracking systems increasingly affordable. Initial evaluations of one of the systems indicates that with careful setup, the accuracy and precision of these systems is comparable to high-cost systems (Ooms, Dupont, Lapon, & Popelka, 2015). The accessibility of these new systems may open the door to researchers interested in exploring the use of eye-based measures as a tool for assessing SA.

The experiment described above analyzed fixation data and SA probe performance post hoc. That is, the probes occurred at pre-assigned times and the eye tracking data before the probe was correlated with probe performance. Another approach to validating eye gaze data as a predictor of SA would be to continuously assess SA with eye tracking and to trigger SA probes at times when either high SA or low SA were predicted by the gaze data. Although the work described here and by others compared eye tracking with performance on an SA probe, there are other SA assessment techniques that may benefit from eye tracking. For example, combined with in-task queries (i.e., SPAM), eye tracking may provide new information about where individuals direct their attention following a query, which would offer additional performance measures and generally inform understanding of a user's SA.

Conclusions

Although traditional measures of SA (e.g., freeze probes, in-task queries, and subjective questionnaires) offer high face validity, these techniques can often be disruptive, difficult to implement, and provide data on SA only at discrete points in time. The use of eye gaze as a non-obtrusive, continuous measure of SA does not suffer from these shortcomings and therefore should be further explored as a complementary means of assessing SA. Furthermore, the use of gaze-based SA measures could provide great value in applied settings, for example in triggering adaptive automation: for providing tailored alerting (Ratwani et al., 2010) or in assessing operator performance with new displays or automation. As a caveat, though, there is a distinction between what an individual attends to in the environment and the quality of the information they obtain from the environment, with SA being a representation of the latter (Vidulich, 2003). Furthermore, fixating on a specific visual control is a requirement for gathering information from it, yet does not always imply processing the information within the visual scene (Sarter, Mumaw, & Wickens, 2007). Therefore, eye gaze data should not be considered in isolation or as a complete picture of an individual's SA, but rather as one tool for gathering a more complete understanding of an individual's SA and the processes used to acquire it. The work presented here also highlights the importance of specifying how eye tracking metrics are calculated and suggests that researchers may need to consider multiple or variable definitions for basic units such as fixations. Research on linking eye gaze and SA is only in its early stages; however, the results appear promising and the introduction of new low-cost equipment should permit new and exciting expansions of this work going forward.

References

Adams, M. J., Tenney, Y. J., & Pew, R. W. (1995). Situation awareness and the cognitive management of complex systems. *Human Factors, 37*, 85–104.

Ahlstrom, U., & Friedman-Berg, F. J. (2006). Using eye movement activity as a correlate of cognitive workload. *International Journal of Industrial Ergonomics, 36*, 623–636.

Bellenkes, A. H., Wickens, C. D., & Kramer, A. F. (1997). Visual scanning and pilot expertise: The role of attentional flexibility and mental model development. *Aviation, Space, and Environmental Medicine, 68*, 569–579.

Berka, C., Levendowski, D. J., Lumicao, M. N., Yau, A., Davis, G., Zivkovic, V. T., . . . Craven, P. L. (2007). EEG correlates of task engagement and mental workload in vigilance, learning, and memory tasks. *Aviation, Space, and Environmental Medicine, 78* (Supplement 1), B231–B244.

Boussemart, Y., & Cummings, M. (2008). *Behavioral recognition and prediction of an operator supervising multiple heterogeneous unmanned vehicles*. Paper presented at the Humans Operating Unmanned Systems, Brest, France.

Calhoun, G. L., Draper, M. H., & Ruff, H. A. (2009). *Effect of level of automation on unmanned aerial vehicle routing task.* In Proceedings of the Human Factors and Ergonomics Society 53rd Annual Meeting, (pp. 197–201). Santa Monica, CA: Human Factors and Ergonomics Society.

Clark, P. J., & Evans, F. C. (1954). Distance to nearest neighbor as a measure of spatial relationships in populations. *Ecology, 35*, 445–453.

Collewijn, H., & Kowler, E. (2008). The significance of microsaccades for vision and oculomotor control. *Journal of Vision, 8*, 1–20.

Crundall, D., Underwood, G., & Chapman, P. (1999). Driving experience and the functional field of view. *Perception, 28*, 1075–1087.

Cummings, M. L., Mastracchio, C., Thornburg, K. M., & Mkrtchyan, A. (2013). Boredom and distraction in multiple unmanned vehicle supervisory control. *Interacting with Computers, 25*, 34–47.

Department of Defense. (2013). Unmanned systems integrated roadmap: FY2013–2038. Washington, DC.

Di Nocera, F., Camilli, M., & Terenzi, M. (2007). A random glance at the flight deck: Pilots' scanning strategies and the real-time assessment of mental workload. *Journal of Cognitive Engineering and Decision Making, 1*, 271–285.

Di Nocera, F., Terenzi, M., & Camilli, M. (2006). Another look at scanpath: distance to nearest neighbour as a measure of mental workload. In D. de Waard, K. A. Brookhuis, & A. Toffetti (eds.), *Developments in human factors in transportation, design, and evaluation* (pp. 295–303). Maastricht, The Netherlands: Shaker Publishing.

Durso, F. T., & Dattel, A. R. (2004). SPAM: The real-time assessment of SA. In S. Banbury & S. Tremblay (eds.), *A cognitive approach to situation awareness: Theory and application* (Vol. 1, pp. 137–154). Hampshire, UK: Ashgate Publishing Ltd.

Endsley, M. R. (1988a). *Design and evaluation for situation awareness enhancement.* In Proceedings of the Human Factors and Ergonomics Society 32nd Annual Meeting, (pp. 97–101). Santa Monica, CA: Human Factors and Ergonomics Society.

Endsley, M. R. (1988b). *Situation awareness global assessment technique (SAGAT).* In Proceedings of the National Aerospace and Electronics Conference, (pp. 789–795). New York: Institute of Electrical and Electronics Engineers.

Endsley, M. R. (1995). Toward a theory of situation awareness in dynamic systems. *Human Factors, 37*, 32–64.

Endsley, M. R. (1999). Situation awareness in aviation systems. In D. J. Garland, J. A. Wise, & V. D. Hopkin (eds.), *Handbook of aviation human factors* (pp. 257–276). Mahwah, NJ: Lawrence Erlbaum Associates.

Ericsson, K. A., & Kintsch, W. (1995). Long-term working memory. *Psychological Review, 102*, 211–245.

Gartenberg, D., Breslow, L., McCurry, J. M., & Trafton, J. G. (2014). Situation awareness recovery. *Human Factors, 56*, 710–727.

Gegenfurtner, A., Lehtinen, E., & Säljö, R. (2011). Expertise differences in the comprehension of visualizations: A meta-analysis of eye-tracking research in professional domains. *Educational Psychology Review, 23*, 523–552.

Gertler, J. (2012). *US unmanned aerial systems* (Report No. 7–5700). Retrieved from http://oai.dtic.mil/oai/oai?verb=getRecord&metadataPrefix=html&identifier=ADA566235.

Holmqvist, K., Nyström, M., Andersson, R., Dewhurst, R., Jarodzka, H., & Van de Weijer, J. (2011). *Eye tracking: A comprehensive guide to methods and measures*. Oxford, UK: Oxford University Press.

Jones, D. G. (2000). Subjective measures of situation awareness. In M. R. Endsley & D. J. Garland (eds.), *Situation awareness analysis and measurement* (pp. 113–128). London: Routledge.

Klein, G. (1997). The recognition-primed decision (RPD) model: Looking back, looking forward. In C. E. Zsambo & G. Klein (eds.), *Naturalistic decision making* (pp. 285–292). New York: Routledge.

Kohlmorgen, J., Dornhege, G., Braun, M., Blankertz, B., Müller, K.-R., Curio, G., Hageman, K., Bruns, A., Schrauf, M., Kincses, W. (2007). Improving human performance in a real operating environment through real-time mental workload detection. In G. Dornhege, J. R. Millan, T. Hinterberger, D. McFarland, & K.-R. Müller (eds.), *Toward brain-computer interfacing* (pp. 409–422). Cambridge, MA: The MIT Press.

Lathan, C. E., & Tracey, M. (2002). The effects of operator spatial perception and sensory feedback on human-robot teleoperation performance. *Presence: Teleoperators and Virtual Environments, 11*, 368–377.

Mackworth, N. (1948). The breakdown of vigilance durning prolonged visual search. *Quarterly Journal of Experimental Psychology, 1*, 6–21.

Moore, K., & Gugerty, L. (2010). *Development of a novel measure of situation awareness: The case for eye movement analysis*. In Proceedings of the Human Factors and Ergonomics Society 54th Annual Meeting, (pp. 1650–1654). Santa Monica, CA: Human Factors and Ergonomics Society.

Ooms, K., Dupont, L., Lapon, L., & Popelka, S. (2015). Accuracy and precision of fixation locations recorded with the low-cost Eye Tribe tracker in different experimental set-ups. *Journal of Eye Movement Research, 8*, 1–24.

Parasuraman, R. (1987). Human-computer monitoring. *Human Factors, 29*, 695–706.

Parasuraman, R., Molloy, R., & Singh, I. L. (1993). Performance consequences of automation-induced "complacency". *The International Journal of Aviation Psychology, 3*, 1–23.

Parasuraman, R., & Riley, V. (1997). Humans and automation: Use, misuse, disuse, abuse. *Human Factors, 39*, 230–253.

Pincus, S. M. (1991). Approximate entropy as a measure of system complexity. *Proceedings of the National Academy of Sciences, 88*, 2297–2301. Washington, DC: National Academy of Sciences.

Ratwani, R. M., McCurry, J. M., & Trafton, J. G. (2010). *Single operator, multiple robots: an eye movement based theoretic model of operator situation awareness*. In Proceedings of the 5th ACM/IEEE international conference on Human-robot interaction (pp. 235–242). New York: Association for Computing Machinery.

Rayner, K. (2009). Eye movements and attention in reading, scene perception, and visual search. *The Quarterly Journal of Experimental Psychology, 62*, 1457–1506.

Salmon, P., Stanton, N., Walker, G., & Green, D. (2006). Situation awareness measurement: A review of applicability for C4i environments. *Applied Ergonomics, 37*, 225–238.

Salmon, P., Stanton, N. A., Walker, G. H., Jenkins, D., Ladva, D., Rafferty, L., & Young, M. (2009). Measuring situation awareness in complex systems: Comparison of measures study. *International Journal of Industrial Ergonomics, 39*, 490–500.

Salvucci, D. D., & Goldberg, J. H. (2000). *Identifying fixations and saccades in eye-tracking protocols*. In Proceedings of the 2000 Symposium on Eye Tracking Research & Applications (pp. 71–78), New York: Association for Computing Machinery.

Sarter, N. B., Mumaw, R. J., & Wickens, C. D. (2007). Pilots' monitoring strategies and performance on automated flight decks: An empirical study combining behavioral and eye-tracking data. *Human Factors, 49*, 347–357.

Schriver, A. T., Morrow, D. G., Wickens, C. D., & Talleur, D. A. (2008). Expertise differences in attentional strategies related to pilot decision making. *Human Factors, 50*, 864–878.

Shic, F., Scassellati, B., & Chawarska, K. (2008). *The incomplete fixation measure*. In Proceedings of the 2008 Symposium on Eye Tracking Research & Applications (pp. 111–114). New York: Association for Computing Machinery.

Taylor, R. (1990). Situational awareness rating technique (SART): The development of a tool for aircrew systems design *Situational Awareness in Aerospace Operations (AGARD-CP-478)* (pp. 3/1–3/17). Neuilly Sur Seine, France.

Teichner, W. H. (1974). The detection of a simple visual signal as a function of time of watch. *Human Factors, 16*, 339–352.

Van de Merwe, K., Van Dijk, H., & Zon, R. (2012). Eye movements as an indicator of situation awareness in a flight simulator experiment. *The International Journal of Aviation Psychology, 22*, 78–95.

Vidulich, M. A. (2003). Mental workload and situation awareness: Essential concepts for aviation psychology practice. In P. Tsang & M. A. Vidulich (eds.), *Principles and practice of aviation psychology* (pp. 115–146). Mahwah, NJ: Lawrence Erlbaum Associates.

Warm, J. S., Parasuraman, R., & Matthews, G. (2008). Vigilance requires hard mental work and is stressful. *Human Factors, 50*, 433–441.

12 Four design choices for haptic shared control

M. M. (René) van Paassen, Rolf P. Boink, David A. Abbink, Mark Mulder, and Max Mulder

Interest for the application of haptic interfaces (or haptic displays) for vehicles is increasing. These interfaces use an operator's sense of feeling or touch to display information about the environment or about the device that is being operated. Nissan Motor Company, Ltd., for example, markets a haptic gas pedal that can provide force feedback to the driver about obstacles or vehicles detected in front of one's car (Mulder, Abbink, Van Paassen, & Mulder, 2011). In aviation, research has been performed on unmanned aerial vehicle (UAV) tele-operation and in-aircraft haptic feedback (De Stigter, Mulder, & Van Paassen, 2007; Lam, Mulder, Van Paassen, Mulder, & Van der Helm, 2009; Goodrich, Schutte, & Williams, 2011). When the forces created by the haptic display can influence the input to the controlled system, a *shared control* situation is created. Both the human operator and the system's automation, through the haptic interface, exert an influence on the control input.

The advantages of haptic shared control over conventional assistance by automation have been argued previously (Abbink, Mulder, & Boer, 2012): since the actions of the automation are presented through the control device, operators are continuously aware of these actions without further taxing the often overloaded visual channel. Additionally, operators can use their quick and adaptable neuromuscular system to respond and easily overrule the automation when necessary. There is substantial evidence that haptic shared control inherits part of the automation benefits—such as improved performance and reduced effort—but also some of the pitfalls of automation (Mars, Mathieu, & Hoc, 2014; Petermeijer, Abbink, & De Winter, 2014). In other words, the reliability of the automation, complacency, (over-)reliance, transparency, and level of automation remain relevant issues, depending on the design choices of shared control. In addition to that, new aspects in shared control are:

1. The continuously variable balance between the human operator's and the automated controller's contributions.
2. The fact that the control input is now the sum of the individual inputs of human and automation.

An implication from the first aspect is that the authority of automation versus the authority of the human operator must be made explicit in the design of the haptic device. The choice for device parameters, such as damping, mass and stiffness, and the tuning of the device's force feedback, all affect this balance. Earlier work in our group explored two design choices for haptic guidance: force feedback and stiffness feedback (Abbink & Mulder, 2009). An important complicating factor in design and evaluation is the human adaptability. Because the human's neuromuscular system can also adapt to haptic settings, this means that this tuning is not easily optimized by trial and error. A large range of haptic tuning settings produce acceptable human-in-the-loop behavior for nominal conditions because humans will adapt their neuromuscular system to generate desired control behavior. Therefore, selection of parameters should be based on a design logic that considers both dynamic properties of the neuromuscular system and task requirements. This issue was explored for car driving and for a system with haptic feedback for a UAV (Abbink et al., 2012; Sunil, Smisek, Van Paassen, & Mulder, 2014).

The movement of the input device, for example a steering wheel, is a result of the sum of the control forces from the human and from the haptic support system. This leads to the unusual situation that two controllers (a human and a device) are active in parallel, rather than in a cascade control, as is common in supervisory control. This means that shared control situations need to be analyzed with respect to their control properties. The control actions by the human and automation must ideally complement each other, and not counteract unnecessarily.

Haptic shared control might be likened to a situation that many student pilots encounter when they learn to fly, typically the instructor's and student's controls are mechanically linked, initially the instructor acts as a supportive controller. The instructor, or a haptic support system, can adapt or be tuned along multiple dimensions. First there is the question of what the *reference behavior* should be. In a flare, for example, should the path be an optimal one, optimizing, e.g., energy use or normal load, or would it be better to follow a workable strategy that is somehow visible and makes sense to the student pilot? Another choice must be made in the *feedback control strategy*, which determines how deviations from the reference path are corrected, either strongly and quickly, or in a more relaxed manner. Most control strategies not only involve corrections from a reference, but also use *feed-forward* control actions that anticipate upcoming maneuvers. The instructor can vary between not using feed-forward and waiting for the student to initiate the maneuver, or providing partial or full feed-forward, essentially contributing partially or fully to the implementation of an upcoming maneuver. A fourth question in the haptic collaboration between two agents is their relative level of authority, a strong and muscular instructor firmly gripping the controls—independent of the setting of feedback or feed-forward strategies—will affect the balance of the *authority* between the two agents controlling the aircraft.

We explored the issue of reference behavior in an automobile driving experiment on curve negotiation with haptic support. It is well known that drivers do not follow the center of the road in corners, but slightly "cut" the corners resulting in better driving comfort, and individual differences exist in this behavior, making the issue of a reference path compatible with human behavior relevant. For our research, we first fit a guidance model to drivers' natural preferences. We took individual fits (individualized guidance (IG)), and an overall fit as an average for the subject population, the "one-size-fits-all" (OSFA) variant. In a test with the haptic support system, the surprising result was that many subjects rejected the IG in favor of the OSFA variant (Boink, Van Paassen, Mulder, & Abbink, 2014).

This chapter discusses this experiment and investigates the possible causes for its findings. Then it lists an overview of the design considerations for haptic shared control that were discovered after analyzing our results, and continues analyzing haptic shared control in the same context, since lane keeping and curve negotiation combine all challenges (a non-obvious reference signal, feedback, and feed-forward) for a shared haptic control system. An architecture for a haptic system supporting shared control and compatible with the design considerations is presented. This architecture is tested with a simplified model of haptic shared control in car driving.

Haptic shared control

In a vehicle with haptic shared control, both the human operator and an automated agent influence the control input; the human through exerting a torque on the steering wheel, and the automation through an additional torque on the wheel from, e.g., an electric motor. In addition, the wheel may have its own dynamic characteristics, typically mass and damping, and the torques on the front wheels (self-aligning torque) are passed through the linkage, resulting in an apparent stiffness of the steering wheel (K_w). Figure 12.1 depicts this situation. This resembles the set-up described in Griffiths and Gillespie (2004, figure 2), which has a slightly different format for the block diagram, since it expressly shows how the self-aligning torque in a simulation is implemented by the electric motor used in that experiment.

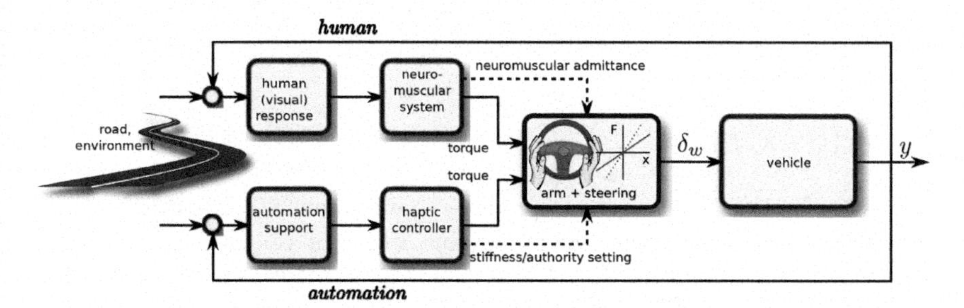

Figure 12.1 Schematic representation of haptic shared control.

Note. Both the automation and human user provide an input torque, which, through the combined dynamics of hands and steering wheel, results in an input (δ_w) for the vehicle.

When the steering wheel—or another control device—is handled by the human operator, the human's muscular force and the torque from the haptic feedback system act on the combined dynamic properties of that coupled system. A human can generally influence the dynamics of his or her limb, by changing the setting of the neuromuscular system, effectively increasing or decreasing apparent limb stiffness (K_{nms}). The combination of the control device stiffness K_w and the human neuromuscular stiffness K_{nms} determines the control device movements δ_w in response to the torque exerted by the human and the haptic support, illustrated in Figure 12.1 by the lines in the force-movement diagram.

If properly equipped, the haptic device's stiffness (and possibly damping and mass as well) can be modified in an analogous manner to the neuromuscular system settings. Such modifications can serve to shift the balance of the human contribution to the system input versus the haptic automation's contribution (Abbink & Mulder, 2009); this modulation is indicated by the dashed arrows in Figure 12.1.

An important component in the haptic shared control is the generation of the guidance. *Two situations* are generally distinguished. The shared control may have the purpose of avoiding collisions with either moving or stationary obstacles. In that case, the haptic display shows *virtual fixtures*, virtual obstacles and boundaries perceived in the haptic device through repulsive forces, *pushing* the vehicle away from danger zones. In the case of automobile driving, when only one lane is considered the guidance can be continuous, and the goal of the automation can be defined as keeping the vehicle on an "optimal" track. Rather than virtual fixtures that provide repulsive forces to reduce the possibility of collision, a continuous virtual fixture is implemented that generates forces that *pull* the vehicle to a specific target.

For an effective haptic interface, this target should coincide with the driving behavior that a human driver would find acceptable. In curves, assuming a position of the automobile on the center of the road does not reflect how human drivers will negotiate a curve. In our experiment (Boink et al., 2014), we identified the manner in which drivers negotiated a curve and fit this with a simple model that calculates the steering wheel angle, given the difference between the nominal track and the lateral position of the automobile at some look-ahead time T_{LH}:

$$\delta_w(t) = K_\delta E_{T_{LH}}(t) \tag{12.1}$$

Here $E_{T_{LH}}$ is the lateral error of a predicted position of the automobile created by integrating a model with the automobile's current velocity and rotational rate over a prediction time T_{LH}. The K_δ and T_{LH} parameters were identified for each subject individually, and Equation (12.1) was used to create the nominal path for the haptic control. In addition, a version of the controller was tested which used parameters in the center of the parameter space observed for all participants (OSFA, indicated with the circle in Figure 12.2).

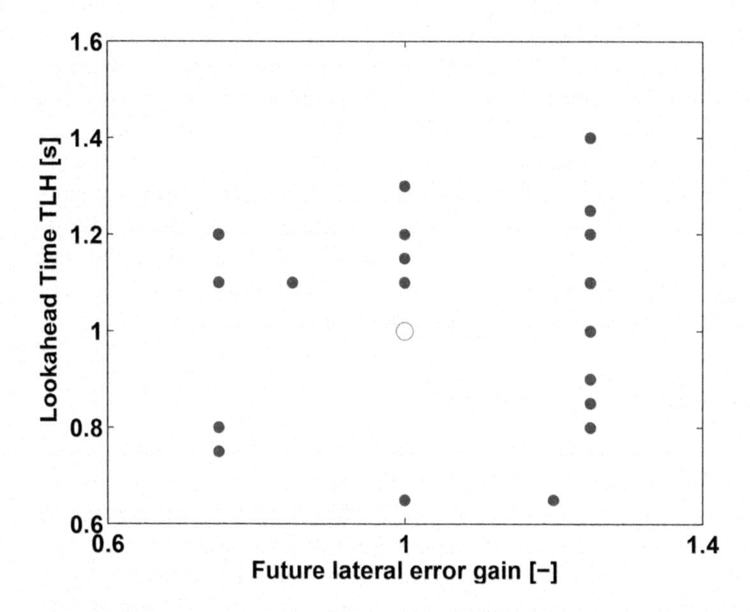

Figure 12.2 Individual fits of look-ahead time and lateral error gain and OSFA choice.

Notes. $K_w K_\delta$, filled circles and OSFA choice (open circle) used in the experiment (Boink et al., 2014).

To convert the nominal path into a guidance force, a scaling gain needs to be determined. In this work, this gain is based on the stiffness of the steering wheel (K_w), on the assumption that torque from the haptic feedback system should generate the proper steering wheel angle when the user does not hold the steering wheel. With this gain, the controller's target position is converted into a torque on the steering wheel:

$$F_h(t) = K_w\left(K_\delta E_{T_{LH}}(t)\right) \tag{12.2}$$

This means that a user holding the steering wheel in the haptic shared control condition should not exert a force on the wheel for a run that is to match exactly the preferred track. Since the wheel still has to move, the only muscular force required is the force needed to overcome the natural stiffness K_{nms} of the user's own neuromuscular system.

Experiment

To test our hypothesis that IG would be preferred over OSFA, an experiment with twenty-four subjects was performed in a fixed-base driving simulator. The simulator was equipped with a Nissan Motor Company, Ltd. steering wheel actuated by a Moog-FCS ECol-8000 S electrical actuator. A visual scene was projected on the walls in front and to the side of the simulator, providing a horizontal field-of-view of almost 180 degrees. In a first session, participants drove a track with alternating left and right curves over 45 degrees, with a 250m radius. No haptic feedback was provided in this session, and the velocity of the simulated automobile was fixed to 80 km/h. An IG model as in Equation (12.1) was fit to the data of this run. On the basis of the IG model fits, also an OSFA fit was determined. In a second session, subjects performed pairs of runs with the two types of haptic guidance, starting with either the IG or the OSFA tuning. Figure 12.2 shows the spread of the tuning parameters and the OSFA tuning. After each pair of runs in the second session, subjects were asked to indicate their preference for either the first or second run.

Experiment results and discussion

Figure 12.3 gives an example of a single curve driven by a "medium-gain" subject, in the OSFA condition. In addition, the curve driven by the subject in the absence of haptic support, and the result of letting the haptic support system "drive alone" is given. A number of surprising results can be noted. First, the lateral error (note that this is the lateral error at the look-ahead point $E_{T_{LH}}$), which is fairly small when the human is driving—with or without haptic support—indicating a successfully driven curve. Second, when the haptic automation drives alone (hands-off steering wheel), the errors are fairly large, indicating that the control law in Equation (12.2) is actually not very effective. Finally, the force from the guidance actually seems to oppose the human torque over a large stretch of the curve. This latter result was found with multiple subjects, and more strongly with subjects with curve negotiation behavior that resulted in model fits with high gains and large look-ahead times.

Initial lessons learned

The experiment addressed one of the main design decisions in creating a haptic support system, namely the question of *what should be the reference trajectory for the haptic support system*, in this case for a system that uses a continuous target reference. Rather than taking the lane's center, which would result in unnatural driving behavior, a simple control law was fitted to previously observed human-only control behavior. Compared to an older experiment (Abbink, Cleij, Mulder, & Van Paassen, 2012) in which a data-based approach was taken whereby the reference was created by averaging a number of previous runs, the present approach is more general. In the context of a haptic support system, we will label this property as the capability to provide a "Human Compatible Reference" (HCR, Figure 12.4).

One of the other surprising results from this experiment was that more subjects preferred the OSFA tuning of the controller over the adaptive tuning. Upon further inspection, it proved that this correlated with the gain of the individual tuning; subjects that had a lower gain preferred the individualized

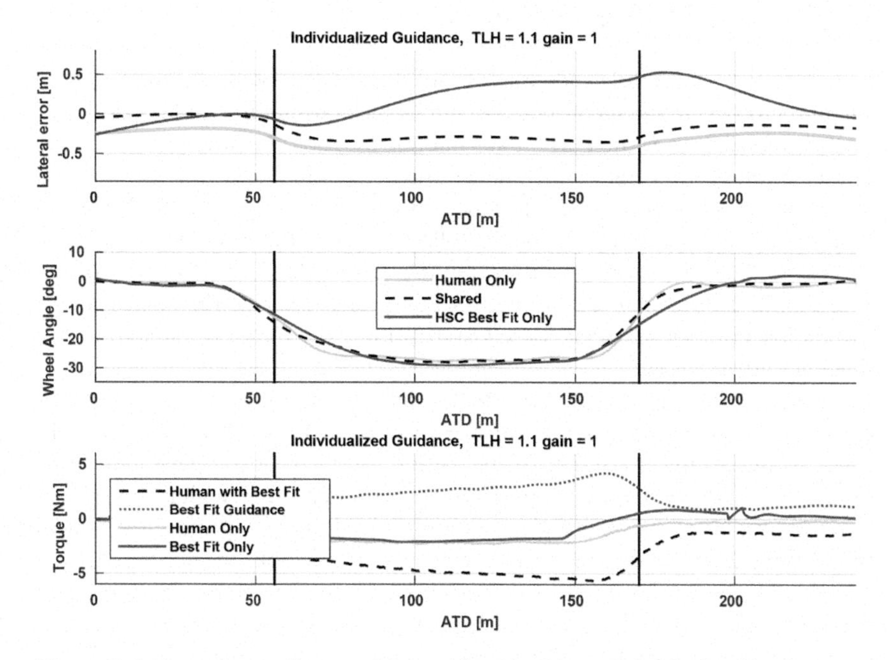

Figure 12.3 Example run from a subject with and without individualized haptic shared control.

Note. Shown negotiating a curve (60–170 m), illustrating in this case initially no, and later an opposite, contribution of the guidance to the steering wheel torque.

tuning settings, and subjects with higher gain preferred the OSFA tuning. To investigate possible causes for this, a further analysis of the controller and the haptic feedback it provides was done.

Given the log data indicate that the haptic guidance often counteracts the human control input, it could be expected that a weaker haptic guidance is to be preferred over a stronger one. This was also found by Mars, Deroo, and Hoc (2014). Their experiments, however, may have forced possibly less natural curve negotiations, since the haptic feedback was based on the lane center, providing an additional reason for subjects to prefer weaker haptic feedback. However, why would the haptic guidance not contribute to the control goal and in many cases counteract the human control? And also, why would haptic guidance alone not create a proper control of the vehicle? These questions will be addressed by means of studies with a simple model of the haptic shared control.

To initially analyze the effect of the control law used in our study, a small-angle approximation can be used for the future lateral position error $E_{T_{LH}}$. The following Equation calculates this, based on the current lateral error ($y(t) - y_r(t)$) and the current heading error ($\Psi(t) - \Psi_r(t)$):

$$\tilde{E}_{T_{LH}} = VT_{LH}(\psi(t) - \psi_r(t)) + y(t) - y_r(t) \tag{12.3}$$

Here V is the automobile's forward speed, T_{LH} is the look-ahead time, generally in the order of one second or less, $\Psi(t)$ is the automobile heading, and $\Psi_r(t)$ is the road heading at the current automobile location. Finally, $y(t)$ is the current position of the vehicle and $y_r(t)$ the road centerline.

The approach used in the experiment was to use a single feedback control loop to provide both the support to follow turns in the reference trajectory, and to provide feedback in case of steering errors; the future position of the automobile is compared to the commanded track (that is, the road center), and the resulting lateral error is used to determine the torque generated by the shared control system.

In the experiment, to calibrate the model for a user's preference in corner cutting, a subject first drove a number of tracks without haptic support. In order to determine the effect of the design tested in our experiment, consider a run with the haptic support system switched on in which the subject exactly generates the steering commands matching the preference model (Equation (12.1)). In that case, the lateral error at the look-ahead point ($E_{T_{LH}}$) is minimal; the only source of deviation between the reference model and the user's run would be the remaining variation in the user's driving that could not be captured by the model. According to Equation (12.2), the haptic feedback force would be nearly zero in this case. Inspection of an experiment with similar architecture in literature (Griffiths & Gillespie, 2004)

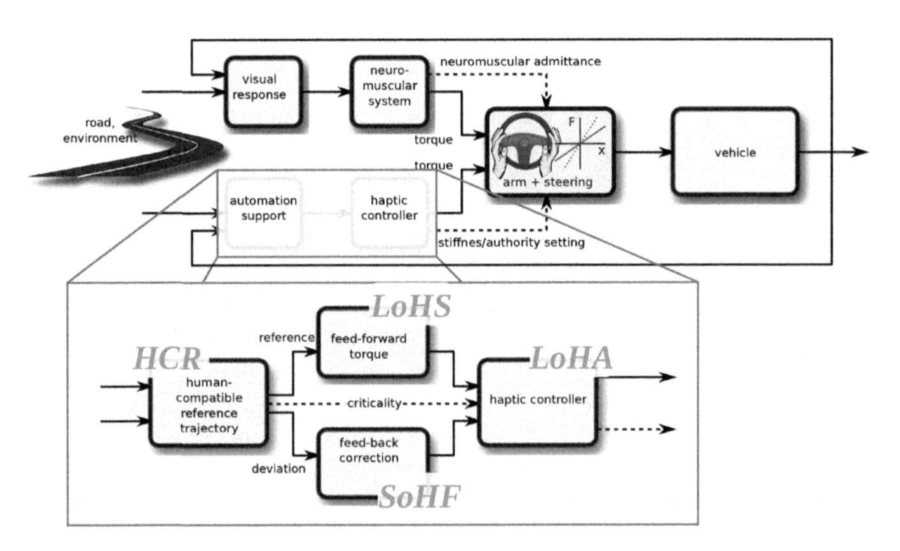

Figure 12.4 Schematic representation of haptic shared control, proposing four components related to the four design choices identified in this chapter.

Notes. HCR = human compatible reference; LoHS = level of haptic support; SoHF = strength of haptic feedback; LoHA = level of haptic authority.

suggests that the same occurred in that set-up; with successful control by the human, and a zero control error, there is no torque contribution from the haptic support for curve negotiation!

This means that to a perfect driver, who perfectly replicates his or her previous input, this haptic support system will not be giving any support at all; only when a later run deviates from the nominal driving behavior will the haptic support introduce corrective forces.

This relates to a second design decision that needs to be made, *how much should the haptic support system contribute to the control effort in nominal (no-deviation from target) cases*? The haptic support system tested in our experiment relied on error between the determined nominal path and the driven path. In that case, subjects who seek support from the haptic system need to first allow deviations from the nominal path—over those used in corner cutting—before getting this support. The hands-free runs in Figure 12.3 are an illustration of this point. Another design choice is to be made for a haptic support system. Since this relates to support in the primary task, it is labeled as "level of haptic support" (LoHS). A low LoHS then corresponds to a system where the user provides the control input for following the target (the road in this case), and at higher levels of haptic support the haptic system adds input to perform this task.

A further explanation for the preference of the OSFA tuning is provided by considering an initial lateral error in the automobile position $y(t)$. For each unit in lateral deviation, a feedback force of $K_w K_\delta (y_r(t) - y(t))$ Nm will be generated. An initial deviation in automobile heading has a similar effect, now with a gain of $K_w K_\delta V T_{LH} (\Psi_r(t) - \Psi(t))$. Results of these calculations with a simple model for the automobile and the haptic support system are given in Figure 12.5. Higher feedback gains $K_w K_\delta$ yield a larger reaction to the initial heading and lateral offset; a larger look-ahead time T_{LH} yields a stronger response to heading errors.

It is clear that the tuning of the control strategy, which was needed to describe each individual driver's track preference, in this architecture *also* affects the strength of the haptic feedback in the presence of a deviation from the ideal path. In the current setting, this error correction was experienced as too strong, and an alternative architecture of the haptic shared control law should thus be used, one that decouples the tuning of the error correction gain from the implementation of the support for following the future trajectory. This relates to a third design decision for the haptic support system, *what should be the strength of feedback to deviations from the nominal path*? This design decision is labeled the choice on the "strength of haptic feedback" (SoHF).

The need to separate the feedback to error from the feed-forward signals needed to follow the upcoming path coincides with Rasmussen's observation (1983) that most human performance is based on feed-forward control, relying on an internal model and preview of the path to generate largely open-loop control signals. Control systems for technical processes are, however, mostly designed as closed-loop systems, without a feed-forward component. Likewise, human control has been more extensively studied for control situations in which closed-loop control tasks are performed with compensatory

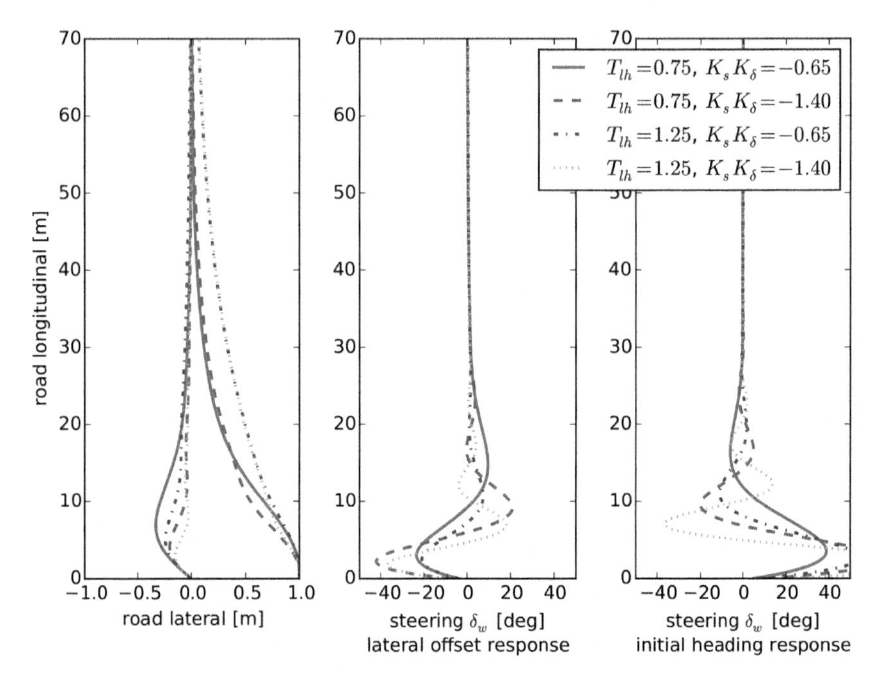

Figure 12.5 Simulation of lateral position and steering wheel responses to an initial offset of 1 m to the right, or 5 degree heading error to the left, for different settings of the $K_w K_\delta$ and T_{LH} tuning.

Note. The left plot shows the lateral position for all conditions (lateral offset and heading error); the middle plot shows δ_w in response to the offset; the rightmost plot shows the δ_w response to the heading error.

(only showing the error between a target and the system's output) displays (McRuer & Jex, 1967). Target following in which the goal is presented has been investigated less extensively, and control with preview of the upcoming target even less so (Van der El, Pool, Damveld, van Paassen, & Mulder, 2016). Haptic support for following a future path thus requires the separation of the haptic support system into two control loops, one based on open-loop control, which provides haptic support to implement the control input for path following, and an additional closed-loop control system, which provides haptic support to minimize deviations from the nominal path. In their implementation, these control loops can be merged, but for design, they should be separated, each obeying their individual requirements.

The final design decision concerns the *authority of the haptic controller*. In analogy to the level of authority in supervisory control (Parasuraman, Sheridan, & Wickens, 2000; Sheridan & Parasuraman, 2005), Abbink et al. (2012) coined the phrase "level of haptic authority" (LoHA). By selecting the control device's stiffness settings—fixed or possibly variable—one can influence the weight of the automation in determining the final control input. A stiff control device will result in a strong weight on the contribution from the haptic support system, and diminish the possible contribution from the human operator. The human operator can counteract this trend by adjusting the settings of his or her neuromuscular system to in turn increase its stiffness, often at the cost of increased effort. An increased stiffness of the neuromuscular system will result in a higher authority for the human; the effect of haptic support forces becomes smaller. The adaptability of the human neuromuscular system means that the combined system is often fairly insensitive to variations in stiffness of the manipulator.

Note that with a high level of authority for the haptic support system, in a system using only feedback control, the haptic interface will still push the system toward the reference. However, before that happens, a control error needs to build up in the system, and the path that results will no longer match the subject's curve negotiation strategy. This behavior annoyed some of the subjects in our experiments; the haptic support basically did not help them, it only penalized deviations.

The following section explores these design decisions and proposes a possible architecture for the haptic support through a simple model of the haptic shared control.

Model studies

Vehicle model

A simplified automobile model, with a model for the wheel dynamics is used to analyze the requirements for shared haptic control systems. It is assumed that combination of the neuromuscular system (thus the driver's arm, muscles, and control) and steering wheel dynamics can be represented by a lumped second-order model. In this model, the total inertia is equal to the sum of the inertia of the steering wheel (I_w) and the effective inertia of the driver's hands (I_{nms}). Likewise, the stiffness and the damping describing this second-order model are equal to the combined stiffness and damping of the steering wheel and hands (K_w, K_{nms}, B_w, B_{nms}, I_w, I_{nms} respectively). This can be described by the following transfer function (a differential equation expressed in the Laplace domain):

$$H_{w+nms}(s) = \frac{1}{(I_w + I_{nms})s^2 + (B_w + B_{nms})s + (K_w + K_{nms})} \tag{12.4}$$

This transfer function gives the resulting steering wheel angle δ_w, in response to the sum of torque from the human's neuromuscular system (F_{nms}; due to muscle activation) and the torque by a haptic assistance system (F_h).

The speed of the simulated automobile is kept constant. It is assumed that the dynamics of tire slip can be neglected; to compensate for slip, an effective ratio between steering wheel and front wheel angle $K_{eff} = 1/15$ is assumed, and the wheel base is $L_{wb} = 2.8$ m. The heading rate $\dot{\psi}$ of the automobile can then be calculated from:

$$\dot{\Psi} = \frac{K_{eff} V}{L_{wb}} \delta_w \tag{12.5}$$

With V being the constant forward speed of the automobile, 22 m/s in this example. The inputs for this model are a torque from the haptic support F_h and a torque from the neuromuscular system F_{nms}. The output is the resulting travel defined as a pair of x, y coordinates, and the automobile's heading.

Control strategy

The control strategy used in the experiment was based on feedback of the lateral position of the vehicle at a projected future point. However, this means that a driver's preference in negotiating a corner is inadvertently coupled to the strength of the feedback when correcting deviations from the planned track, as was argued before and illustrated in Figure 12.5.

An alternative control strategy is therefore needed. If the haptic support system needs to supply a contribution to the steering input that is needed for following the curves in the road, it needs separate information about the road only; in other words, it needs the "target" signal for this. This can then be fed into a feed-forward control law that takes one input signal (the road track), and produces as output the needed haptic assistance torque on the steering wheel.

The road in this case can be defined by the curvature as a function of longitudinal coordinate, $\kappa(l)$, with constant speed, the distance traveled is $l = Vt$. For perfectly following the centerline, the turn rate of the automobile at a constant forward speed V should match the road curvature as follows:

$$r(t) = \kappa(tV)V \tag{12.6}$$

Assuming that wheel slip has been accounted for by using an effective ratio for the rotation of wheel due to steering rotation K_{eff}, the needed steering angle can thus be calculated from the target turn rate as:

$$\delta_w(t) = r(t)\frac{L_{wb}}{K_{eff}} \tag{12.7}$$

If the dynamics of the steering wheel and neuromuscular system can be neglected (or compensated for), the required haptic torque can be calculated by multiplying the required steering angle with the steering wheel stiffness K_w; applying that torque leads to the required equilibrium steering wheel angle. The above steering law will generate inputs that track the road's center line, without any corner cutting, which will thus feel unnatural to most human drivers.

For the aforementioned experiment, we used a closed-loop steering control law, with lateral deviation as input, to generate the haptic force. The control law was tuned to implement corner cutting. However, now we seek an open-loop control law that can generate the haptic force solely on the data from the road, without reference to the vehicle's current lateral position or heading. It should display similar characteristics as observed in human driving, thus starting a turn before the actual corner starts, slightly keeping to the inside of the corner (cutting), and smoothly ending the turn again. As a basis, we can start with Equation (12.7), and to implement corner cutting, instead use the curvature at a road position slightly ahead of the vehicle, so steering commences before the start of a turn. Assuming a look-ahead time $T_{LH_{hs}}$, then at time t the target turn rate would be $\kappa((t + T_{LH_{hs}})V)$.

Without further modification, such a control law would produce abrupt changes in steering wheel angle and rotation rate, and enter as well as end a turn too early. To make the signal more fluent, a filter is applied to this target curvature. In this model a second-order filter is considered sufficient, leading to the following transfer function describing the filtered target in response to the raw target turn curvature:

$$H_{f_{hs}}(s) = \frac{1}{\left(1 + \tau_{f_{hs}} s\right)^2} \tag{12.8}$$

Applying the transfer function makes the input for the haptic support and thereby the steering smoother. The parameter $\tau_{f_{hs}}$ is a filter constant; the larger it is chosen, the smoother the resulting signal. Using the automobile parameters (wheel base L_{wb}, velocity V, effective steering ratio K_{eff} and the steering wheel stiffness K_w), the haptic support torque can then be calculated from the filtered target turn curvature as follows:

$$F_{hs}(t) = K_{LoHS} \frac{K_w L_{wb} V}{K_{eff}} K_{f_{hs}}(t) \tag{12.9}$$

Here K_{LoHS} is the parameter that determines how much the haptic system will contribute to the control action needed to follow the track; when $K_{LoHS} = 1$, the haptic support system will provide hands-free implementation of the control. An example of the generated torque from this filter for negotiating a single turn is given in Figure 12.6. Due to its filtering nature, this control law slightly delays the commanded haptic support torque (continuous line) with respect to the road curvature at the look-ahead point on which the calculation is based. For proper behavior of the support system, the look-ahead time $T_{LH_{hs}}$ and the filter constant $\tau_{f_{hs}}$ need to be matched, considering the dynamics of the steering wheel and the vehicle, to ensure that turns are entered not too late nor too early.

A second function of the haptic support system would be to assist in the correction of deviations or errors from the track. This is then based on a feedback control law. The error signal should in such a case be defined as

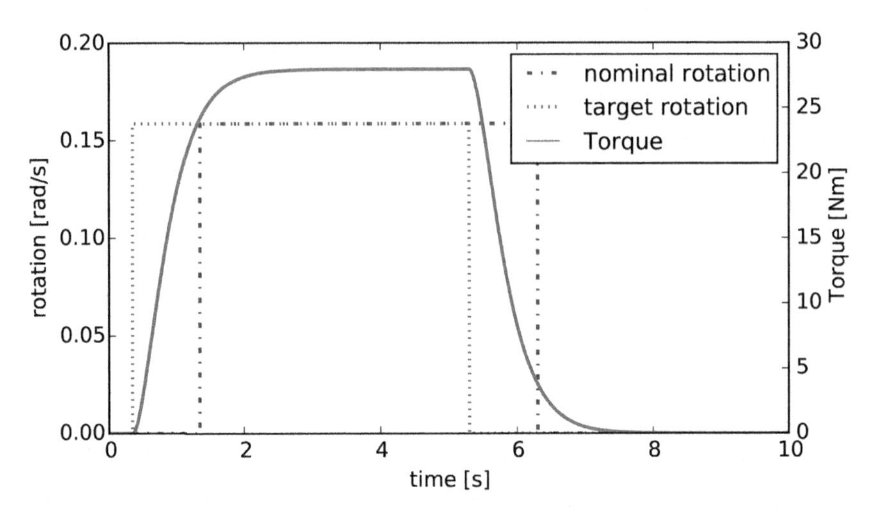

Figure 12.6 Calculation of the feed-forward torque signal for a curve over 45 degree with a length of 110 m.

Notes. Parameters for this simulation $T_{LH_{hs}}$ = 1.0 s, $\tau_{f_{hs}}$ = 0.285 s, K_w = 4.2 Nm·rad^{-1} and L_{wb} = 2.8 m. Note that torque is displayed on a second axis.

the difference between the realized steering actions and the steering actions from a "perfect" run with only the feed-forward control. Perfect in this case would need to include the human driver's corner cutting behavior, otherwise the feed-forward control law and the feedback law interfere and force a non-natural control strategy on the driver. With this in mind, a control structure for the haptic support system uses a prediction for the lateral position with respect to the center line generated by the block calculating the signal for the haptic feed-forward torque.

The look-ahead time for the haptic support, in combination with the automobile speed, determines by how much a corner is cut. If in negotiating the corner, a point at $T_{LH_{hs}}V$ m ahead of the automobile is kept on the center of the lane, one can after Taylor expansion of the resulting goniometric expression of the corner cutting distance, approximate the distance between the automobile and the center line of the lane as follows:

$$y_{cc}(t) \approx \frac{1}{2}\kappa_{f_{hs}}(t)(T_{LH_{hs}}V)^2 \tag{12.10}$$

Here $\kappa_{f_{hs}}(t)$ is again the filtered curvature of the road. A corrective torque for any deviation from this reference can be determined by projecting the future position of the vehicle at $T_{LH_{hs}}$ ahead, y_{pred}, and comparing that with the reference $y_{cc}(t)$ calculated above:

$$F_{hf} = K_{SoHF}(y_r + y_{cc} - y_{pred}) \tag{12.11}$$

The look-ahead time for the feedback can be chosen independently from the look-ahead time for the haptic support; a large look-ahead time places a larger emphasis on heading deviation. The gain factor K_{SoHF} determines the strength of the reaction to lateral deviation, with $y_{T_{lh}}$ and $y_{cc}(t)$ both being defined with respect to the road centerline.

Figure 12.7 shows the effect of the combination of the feed-forward and feedback in the haptic support. This model simulates hands-off control and illustrates addition of the feedback to the open-loop haptic feed-forward system. Figure 12.7 also shows the result of steering on the feed-forward haptic support signal alone; this line shows a small misalignment after the turn due to the simplifying assumptions in the generation of the feed-forward signal. With lateral feedback, the vehicle is brought effectively back on track, with a relatively small correction. This illustrates that complete perfection in the feed-forward path is not needed; it will be impossible to reach that in an actual implementation anyhow.

To illustrate the fact that correct path following is relatively independent from feedback strength, two variations of the feedback control law have been tried; one with a high $K_w K_\delta$ gain of 1.25, and a look-ahead time 1.4 s, and one with the lowest gain and look-ahead, 0.75 and 0.75s.

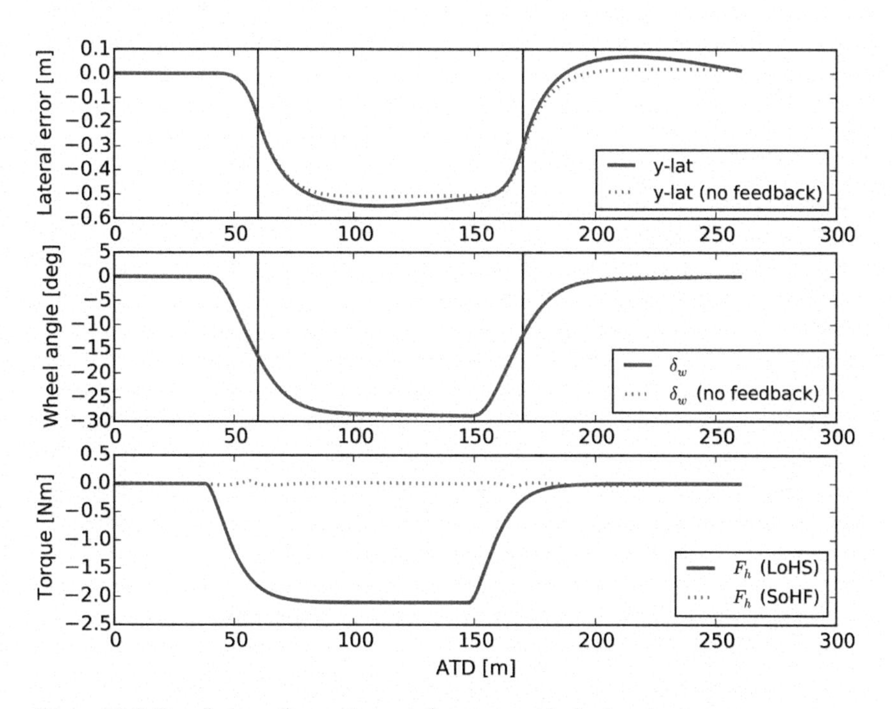

Figure 12.7 Simulation of negotiating left corner with the haptic support system.
Note. For this simulation T_{LHhs} $T_{LH_{hs}}$ = 1.0 s, τf_{hs} = 0.285 s, $T_{LH_{hf}}$ = 0.65 s, $K_w K_\delta$ = 0.42, other parameters and road data are as for the experiment.

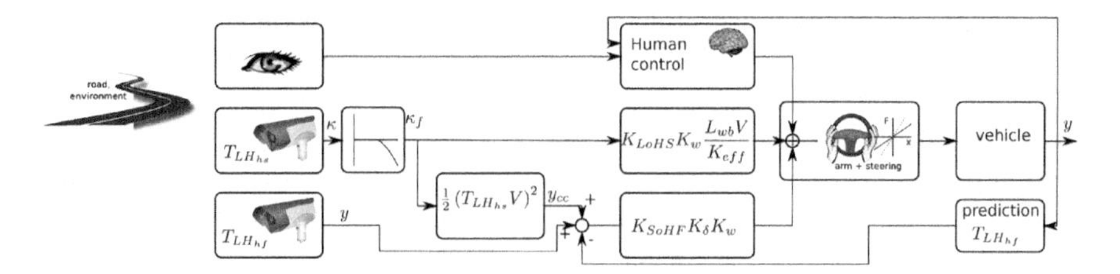

Figure 12.8 Overview of the proposed architecture for the haptic support system.

Note. Road curvature κ forms the basis of haptic support and corner cutting reference values. Preview lateral deviation provides the reference for haptic feedback.

The strategy of combining a haptic feed-forward and a feedback contribution can be illustrated in the haptic architecture by splitting the haptic support system into two separate blocks (Figure 12.8). The first block provides the support for following the target signal; settings for this block, as demonstrated above, determine the LoHS. As a second output, this control block provides the expected track that it calculates for the automobile, including possible adaptations such as corner cutting. The second block provides the support for correcting deviations, on the basis of the expected ideal track from the first block and the measured movement from the vehicle, implementing the SoHF through an independent tuning.

Conclusion

Training settings in aircraft, in which the instructor's and student's controls are mechanically linked, can be seen as a reference model for haptic shared control system. Here the instructor acts as a supportive controller, illustrating desired steering behavior at the operational level. A good instructor can make a student feel the necessary inputs, and reduce his or her LoHS and SoHF as the student progresses in skill—both by generating less or no feed-forward and reducing corrective feedback inputs, and there is a common and compatible (visual frame of) reference, ensuring that the reference is compatible with the student (HCR). As needed, the instructor can balance the haptic authority by modifying his or her grip, providing a variable LoHA. An implementation of haptic shared control with automation requires that such an instructor's behavior be made explicit with a number of design choices:

1 *Human compatible reference (HCR).* Generation of a reference for the control, compatible with user strategies and the device and environment constraints. Note that this reference must be perceivable also through the other senses, so that the human operator can verify the functioning of the haptic support system.

2 *Level of haptic support (LoHS).* A choice for the level of haptic support; that is, by how much will the automated system contribute to implementing a path that follows the reference (feed-forward).
3 *Strength and strategy of haptic feedback (SoHF).* A choice for the strength of the haptic feedback and the control law upon which this feedback is based (in this case, weighing lateral and heading error); that is, by what control law/aggressiveness will the automation provide corrective inputs to reduce the difference between the reference and the vehicle's path. Care should be taken so the support and feedback match each other.
4 *Level of haptic authority (LoHA).* A choice for the level of haptic authority; that is, what is the balance between human input and automation. A high level of authority is implemented by choosing a large base stiffness of the control device. In that case, the feedback and autonomy signals (since they are adapted to the joint control device and human operator stiffness) scale with that stiffness, and it becomes harder to override the haptic support system.

The first and fourth issues have been addressed in literature. Independently tuning the level of haptic support and the strength of the haptic feedback is a step that is still lacking in many designs. This chapter provides the arguments for making these choices, illustrating this through a haptic architecture that offers individual tuning of the level of haptic support and the strength of feedback.

References

Abbink, D. A., Cleij, D., Mulder, M., & Van Paassen, M. M. (2012). The importance of including knowledge of neuromuscular behaviour in haptic shared control. In *2012 IEEE International Conference on Systems, Man, and Cybernetics (SMC)* (pp. 3350–3355). Seoul, Korea: IEEE. doi:10.1109/ICSMC.2012.6378309.

Abbink, D. A., & Mulder, M. (2009). Exploring the dimensions of haptic feedback support in manual control. *ASME Journal of Computing and Information Science in Engineering, Special Issue on Haptics, 9*(March), 1–9.

Abbink, D. A., Mulder, M., & Boer, E. R. (2012). Haptic shared control: Smoothly shifting control authority? *Cognition, Technology & Work, 14*, 19–28.

Boink, R., Van Paassen, M. M., Mulder, M., & Abbink, D. A. (2014). Understanding and reducing conflicts between driver and haptic shared control. In Chen, C. L. P., & Gruver, W. A. (eds.), *IEEE Systems, Man and Cybernetics Conference* (pp. 1529–1534). San Diego, CA: IEEE.

De Stigter, S., Mulder, M., & Van Paassen, M. M. (2007). Design and evaluation of a haptic flight director. *Journal of Guidance, Control, and Dynamics, 30*, 35–46.

Goodrich, K., Schutte, P., & Williams, R. (2011, September). Haptic-multimodal flight control system update. In *11th AIAA Aviation Technology, Integration, and Operations (ATIO) Conference. American Institute of Aeronautics and Astronautics.* Retrieved from http://dx.doi.org/10.2514/6.2011-6984 (accessed March 2, 2015).

Griffiths, P. G., & Gillespie, R. B. (2004). Shared control between human and machine: Haptic display of automation during manual control of vehicle heading. In *Proceedings of the 12th International Symposium on Haptic Interfaces for Virtual Environment and Teleoperator Systems* (Vol. 27, pp. 358–366). Chicago, IL: IEEE.

Lam, T. M., Mulder, M., Van Paassen, M. M., Mulder, J. A., & Van der Helm, F. C. T. (2009). Force–stiffness feedback in uninhabited aerial vehicle teleoperation with time delay. *Journal of Guidance, Control, and Dynamics*, *32*, 821–835.

Mars, F., Deroo, M., & Hoc, J.-M. (2014). Analysis of human-machine cooperation when driving with different degrees of haptic shared control. *IEEE Transactions on Haptics*, 7, 324–333.

McRuer, D. T., & Jex, H. R. (1967). A review of quasi-linear pilot models. *IEEE Transactions on Human Factors in Electronics*, *HFE-8*, 231–249. doi:10.1109/THFE.1967.234304.

Mulder, M., Abbink, D. A., Van Paassen, M. M., & Mulder, M. (2011). Design of a haptic gas pedal for active car-following support. *IEEE Transactions on Intelligent Transportation Systems*, *12*, 268–279.

Parasuraman, R., Sheridan, T. B., & Wickens, C. D. (2000). A model for types and levels of human interaction with automation. *IEEE Transactions on Systems, Man, and Cybernetics. Part A, Systems and Humans*, *30*, 286–297.

Petermeijer, S. M., Abbink, D. A., & De Winter, J. C. F. (2014). Should drivers be operating within an automation-free bandwidth? Evaluating haptic steering support systems with different levels of authority. *Human Factors*, *57*, 5–20.

Rasmussen, J. (1983). Skills, rules, and knowledge: Signals, signs, and symbols, and other distinctions in human performance models. *IEEE Transactions on Systems, Man and Cybernetics*, *SMC-13*, 257–266.

Sheridan, T. B., & Parasuraman, R. (2005). Human-automation interaction. *Reviews of Human Factors and Ergonomics*, *1*, 89–129.

Sunil, E., Smisek, J., Van Paassen, M. M., & Mulder, M. (2014). Validation of a tuning method for haptic shared control using neuromuscular system analysis. In *Proceedings of the IEEE International Conference on Systems, Man and Cybernetics* (pp. 1518–1523). San Diego, CA: IEEE.

Van der El, K., Pool, D. M., Damveld, H. J., Van Paassen, M. M., & Mulder, M. (2016). An empirical human controller model for preview tracking tasks. *IEEE Transactions on Cybernetics*, 1–13, 46, 2609–2621.

Index

Bold page numbers indicate figures, *italic* numbers indicate tables.